Rotating Machinery Research and Development Test Rigs

Maurice L. Adams, Jr.

CRC Press
Taylor & Francis Group
Boca Raton London New York

CRC Press is an imprint of the
Taylor & Francis Group, an **informa** business

CRC Press
Taylor & Francis Group
6000 Broken Sound Parkway NW, Suite 300
Boca Raton, FL 33487-2742

International Standard Book Number-13: 978-1-138-03238-5 (Hardback)

Library of Congress Cataloging-in-Publication Data

Names: Adams, Maurice L., Jr.
Title: Rotating machinery research and development test rigs / Maurice L. Adams.
Description: Boca Raton : Taylor & Francis, CRC Press, [2017] | Includes bibliographical references.
Identifiers: LCCN 2017000357| ISBN 9781138032385 (hardback) | ISBN 9781138032422 (ebook)
Subjects: LCSH: Rotors--Testing. | Testing--Equipment and supplies.
Classification: LCC TJ1058 .A334 2017 | DDC 621.8/2--dc23
LC record available at https://lccn.loc.gov/2017000357

Visit the Taylor & Francis Web site at
http://www.taylorandfrancis.com

and the CRC Press Web site at
http://www.crcpress.com

MIX
Paper from responsible sources
FSC www.fsc.org FSC® C013056

Printed and bound in Great Britain by
TJ International Ltd, Padstow, Cornwall

This book is dedicated to my late parents and late brother

Maury, Libby, and George

And to my late wives

Heidi and Kathy

And to my four mechanical engineering sons

Maury, Professor Dr. Mike, RJ, and Nate

Contents

Preface

It's hard to think of any machinery type that does not have at least one rotating part. Rotating machinery (RM) is at the heart of the modern world, thus so is RM engineering. The technology of RM is a field of study with considerable depth and breadth, utilizing first principles of all the mechanical engineering fundamental disciplines: solid mechanics, dynamics, fluid mechanics, thermodynamics, heat transfer, and controls. Nearly all industries rely heavily on the reliable operation of RM. These industries include (1) power generation; (2) petrochemical; (3) manufacturing; (4) land, sea, and air transportation; (5) heating, ventilating, and air conditioning; (6) aerospace propulsion; (7) farming; (8) computer disk drives; (9) textiles; (10) home appliances; and (11) a wide variety of military systems.

The ever-present competitive pressures to have machines run faster; be more compact, more powerful, more energy efficient; possess higher power-to-weight ratios; and be less costly have fostered a continuous research and development (R&D) history in the field of RM. That has always necessitated the important component of experimental work to augment the theory-based design analyses. However, modern computer-based analysis tools like finite element analysis (FEA) and computational fluid dynamics (CFD) have significantly impacted on the role of machinery component testing. Prior to such modern computer-based analysis tools, testing was often deemed necessary because of inherent approximations and uncertainties in design methods, and the absence of closed-form solutions to many of the governing theory-based equations, for example, Navier–Stokes equations for 3D fluid flow and the 3D elasticity equations for stress, strain, and deformation in solids. Today, testing is often employed to validate the modern computer software, providing the empirical inputs needed for perfecting the accuracy of those computer codes. Consequently, computer-based analyses are now reliably employed to substitute for some of the pre–computer age product R&D and proof testing. The gas turbine jet engine is one RM high-tech product where this modern engineering approach has provided considerable development cost reductions by making feasible significant reductions in some of the costly development testing of new aircraft engine configurations.

To provide a resource detailing several important rotating machinery R&D test facilities, this book is comprised of 19 chapters describing test rigs pertaining to various types of rotating machinery, including (1) large steam turbine generator sets; (2) power plant, slurry, and heart centrifugal pumps; (3) gas turbines; (4) jet engines; (5) bearings, seals, and rotor dynamics; (6) machine tool spindles; and (7) machinery condition monitoring.

Acknowledgments

Truly qualified technologists invariably acknowledge the shoulders upon which they stand. I am unusually fortunate in having worked for several expert-caliber individuals during my 14 formative years of industrial employment prior to becoming a professor in 1977, especially my four years at the Franklin Institute Research Laboratories (FIRL) followed by my six years at the Westinghouse Corporate R&D Center's Mechanics Department. I am also highly appreciative of many subsequent rich interactions with other technologists. I acknowledge here those individuals, many of whom have unfortunately passed away over the years. They were members of a *now extinct breed of giants* who unfortunately have not been replicated in today's industrial workplace environment.

My work in rotating machinery began (1963–1965) at the Allis-Chalmers Hydraulic Products Division in my hometown York, PA. There, I worked on hydroelectric turbine design. That was followed by employment (1965–1967) at Worthington's Advanced Products Division (APD) in Harrison, NJ. There, I worked under two highly capable European-bred engineers, chief engineer Walter K. Jekat (German) and his assistant John P. Naegeli (Swiss). John Naegeli later returned to Switzerland and eventually became general manager of Sulzer's Turbo-Compressor Division and later general manager of their Pump Division. The APD general manager was Igor Karassik, the world's most prolific writer of centrifugal pump articles, papers, and books and an energetic teacher on centrifugal pumps for all the then-young recent engineering graduates at APD, like me. My first assignment at APD was basically to be "thrown into the deep end" of a new turbomachinery development for the U.S. Navy that even today would be considered highly challenging. That new product was comprised of a 42,000 rpm rotor having an overhung centrifugal air compressor impeller at one end and an overhung single-stage impulse steam turbine powering the rotor from the other end, with water-lubricated turbulent fluid-film bearings. Worthington sold several of these units over a period of many years.

I seized upon an opportunity to work (1967–1971) for an internationally recognized group at the FIRL in Philadelphia. I am eternally indebted to several FIRL technologists for the knowledge I gained from them and for their encouragement for me to pursue graduate studies part time, which led to earning my engineering master's degree at a local Penn State extension near Philadelphia. The list of individuals I worked under at FIRL is almost a *who's who list* for the field and includes the following: Elemer Makay (centrifugal pumps), Harry Rippel (fluid-film bearings), John Rumbarger (rolling-element bearings), and Wilbur Shapiro (fluid-film bearings, seals, and rotor dynamics). I also had the privilege of working with a distinguished group of

FIRL's consultants from Columbia University, specifically Professors Dudley D. Fuller, Harold G. Elrod, and Victorio "Reno" Castelli.

My Franklin Institute job gave me the opportunity to publish in my field. That bit of national recognition helped provide my next job opportunity (1971–1977) at what was truly an internationally distinguished industrial research group, the Mechanics Department at Westinghouse's R&D Center near Pittsburgh. The main attraction for accepting that job was my new boss, Dr. Albert A. Raimondi, leader of the bearing mechanics section, whose famous papers on fluid-film bearings are referenced and reproduced in every undergraduate machine design book. An important insight quite relevant to this book that Al Raimondi imparted to me is paraphrased here: *Design of a unique test rig embodies a caveat differentiating it from other machine design efforts. That is, one usually has spent more than half their budget by the first time they try to operate the rig, and they never work the first time. So when designing a test rig, make a list of all the potential malfunction sources (too numerous to design for all of them) and have a feasible backup redesign modification to correct any couple of these within budget*. In other words, *don't paint yourself into a corner* with the rig design not fixable.

An added bonus at Westinghouse was the presence of the person holding the department manager position, A. C. "Art" Hagg, the company's internationally recognized rotor vibration specialist. My many interactions with Art Hagg were all professionally enriching. At Westinghouse, I was given the lead role on several "cutting-edge" projects, including the nonlinear dynamics of flexible multibearing rotors for large steam turbines and reactor coolant pumps; bearing load determination for vertical multibearing pump rotors; seal development for refrigeration centrifugal compressors; and the turning-gear slow-roll operation of journal bearings, developing both test rigs and new computer codes for these projects. I became the junior member of an elite ad hoc trio that included Al Raimondi and D. V. "Kirk" Wright (manager of dynamics section). They encouraged me to pursue my PhD part time, which I completed at the University of Pittsburgh in early 1977. I express special gratitude to my PhD thesis advisor at Pitt, Professor Andras Szeri, who considerably deepened my understanding of the overlapping topics of fluid dynamics and continuum mechanics.

Since entering academia in 1977, I have benefited from the freedom to publish widely and to apply and extend my accrued experience and knowledge through numerous consulting projects for rotating machinery manufacturers and electric utility companies. I appreciate the many years of support for my funded research provided by the Electric Power Research Institute (EPRI) and the NASA Glenn Research Laboratories.

Academic freedom has also made possible leaves to work abroad with some highly capable European technologists, specifically at the Brown Boveri Company BBC (Baden, Switzerland), Sulzer Pump Division (Winterthur, Switzerland), KSB Pump Company (Frankenthal, Germany), and the Swiss Federal Institute (ETH, Zurich). At BBC, I developed a lasting friendship with my host Dr. Raimund Wohlrab. At the Sulzer Pump Division, I was fortunate to interact with Dr. Dusan Florjancic (engineering director), Dr. Ulrich Bolleter (vibration engineering), and Dr. Johan Guelich (hydraulics engineering). At the KSB Pump Company, I was fortunate to interact with Peter Hergt (Head of KSB's Central Hydraulic R&D, 1975–1988) and his colleagues. I particularly cherish the interactions with my host and dear friend at the Swiss Federal Institute ETH-Zurich, the late Professor Dr. Georg Gyarmathy, the ETH turbomachinery professor, 1984–1998. This book rests upon the shoulders of all whom I have acknowledged here.

Author

Maurice L. Adams, Jr. is founder and past president of Machinery Vibration Inc. (www.mvibe.com) and is professor emeritus of mechanical and aerospace engineering at Case Western Reserve University. Dr. Adams is an author of more than 100 publications and a holder of U.S. patents; he is a life member of the American Society of Mechanical Engineers. He received his BSME (1963) from Lehigh University, MEngSc (1970) from the Pennsylvania State University, and PhD (1977) in mechanical engineering from the University of Pittsburgh. Dr. Adams worked on rotating machinery engineering for 14 years in industry prior to becoming a professor in 1977.

Since then, he has been retained as a rotating machinery consultant by several machinery manufacturers and users in the United States and abroad, including GE Aircraft Engine Group, InVision Technologies, ABB Corporate Research, Rolls-Royce Power Systems, ABB Large Rotating Apparatus, United Technologies Carrier Group, EPRI, Eaton Corporation Manufacturing Technologies Center, Reliance Electric Motors Group, FIRL, Caterpillar Engine Division, Brown-Boveri Large Steam Turbines, Battelle Research, Sulzer Company Pump Division, Oak Ridge National Laboratories, TRW Aerospace Systems, John Deere Tractor Group, and several electric power plants in the United States and abroad.

Dr. Adams has authored two other Taylor & Francis/CRC Press books: *Rotating Machinery Vibration* (2010) and *Power Plant Centrifugal Pumps* (2017). He has been the MS thesis and PhD dissertation advisor to more than 30 graduate students, three of whom are now endowed-chair professors. He was the recipient in 2013 of the Vibration Institute's Jack Frarey Medal for his contributions to the field of rotor dynamics.

1

Radial and Axial Rotor Force Design Technology

Rotor–motion interaction and *unsteady-flow* forces are constantly present within many types of rotating machinery and are the unavoidable inherent nature of these machines. As the author (2010) describes, the characterization of these forces is of utmost importance both at the *design stage* of a machine and in the field when *troubleshooting* problems like *excessive acoustic noise, excessive vibration,* and associated *material fatigue failures*. Figure 1.1 illustrates the author's (2010) major example of a high-energy machine where such forces have a dominant influence on the machine's *durability* and *life expectancy*. As the author further explains there, theoretical engineering-science-based predictions of these forces and their motion gradients (stiffness, damping, and virtual mass) are fraught with considerable uncertainties. Furthermore, their experimental determinations are expensive and can constitute the primary life work of the serious researchers who seek to measure the intensity and complexity of these forces. This chapter highlights the double-spool shaft rig concept as one approach to radial force experimental undertakings and a unique approach to axial rotor force research testing. Competing approaches are explained and evaluated in subsequent chapters.

1.1 Double-Spool Spindle at Case for Fluid-Annulus Vibration

The spindle assembly illustrated in Figures 1.2 through 1.4 was developed by the author for a research sponsored by the Electric Power Research Institute (EPRI) and NASA at the Case School of Engineering on fluid-annulus dynamic forces, that is, hydrodynamic and hydrostatic journal bearings, radial seals, and pump wear-ring fluid annuli (Adams 1983). It is uniquely designed utilizing high-precision ABEC-8 (Fag) angular contact ball bearings sized to withstand spindle bearing static loads up to 25,000 pounds (111,364 N), that is, the large axial loads from the maximum 500 psi (353 N/cm^2) test pressure drop across a boiler feed water pump wear-ring annulus. The spindle bearings are oil-mist lubricated through connecting axial and radial feed holes in the spindle. Not shown in Figure 1.2 is the 20 hp vertical 9-stage centrifugal pump that supplies the pressurized through flow

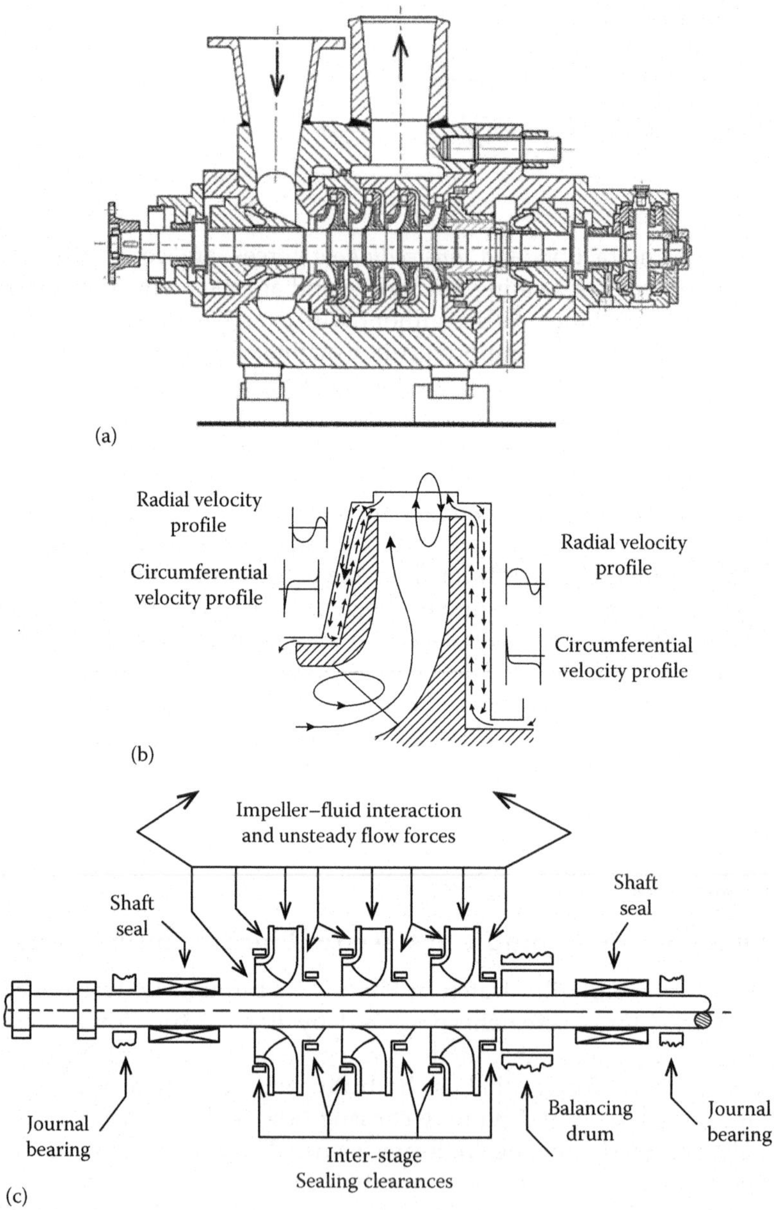

FIGURE 1.1
Multistage feed water pump: (a) cross section, (b) low-flow fluid velocity pattern, and (c) rotor force sources.

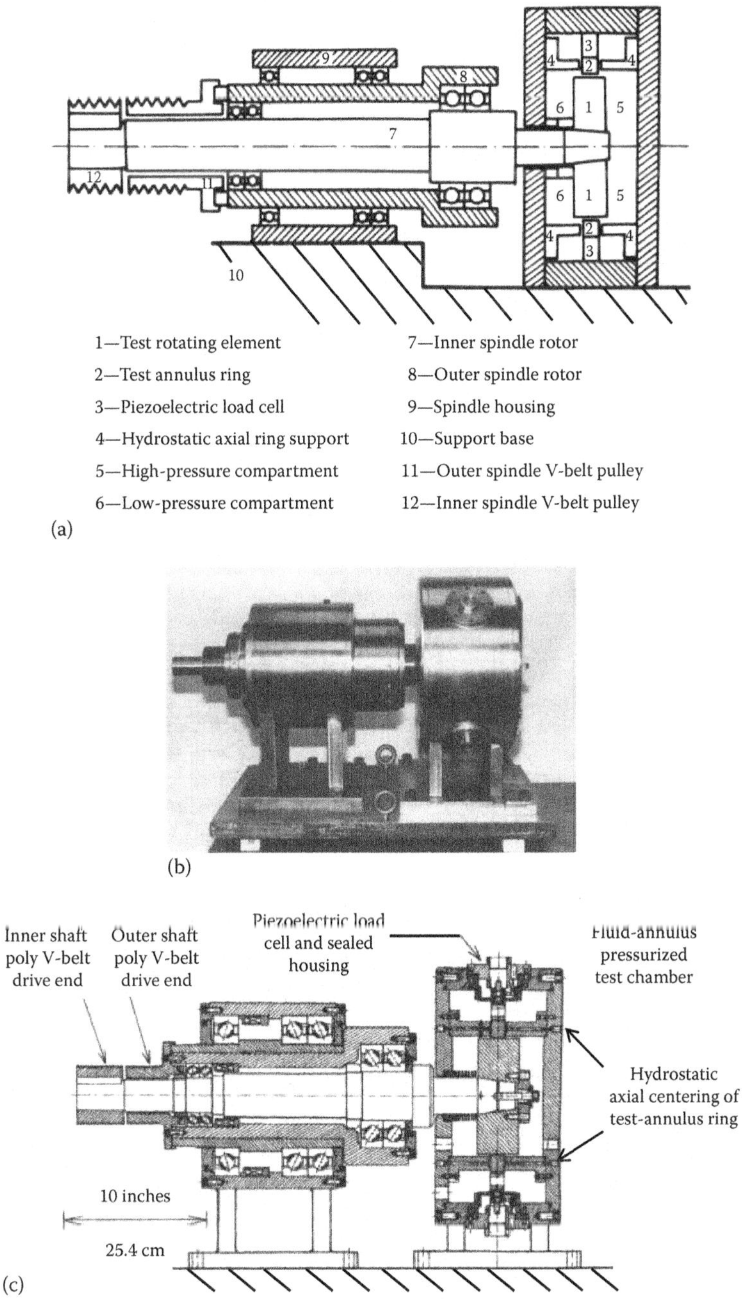

FIGURE 1.2
Double-spool spindle; independent-speed and orbit-frequency control: (a) functional drawing, (b) photo, and (c) cross section.

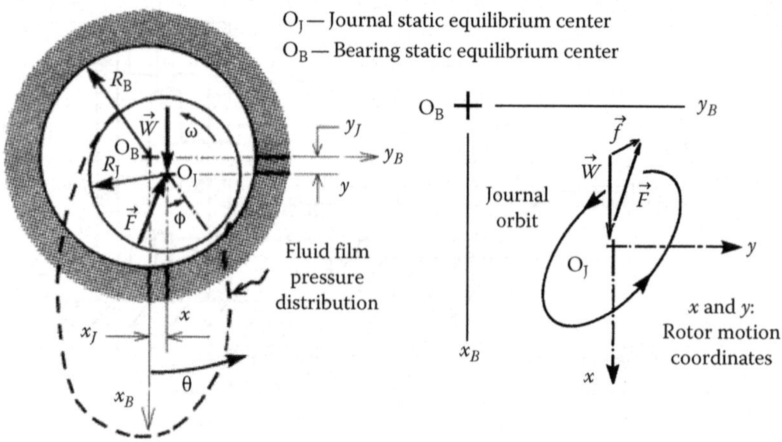

FIGURE 1.3
Journal bearing/liquid annulus configuration and nomenclature.

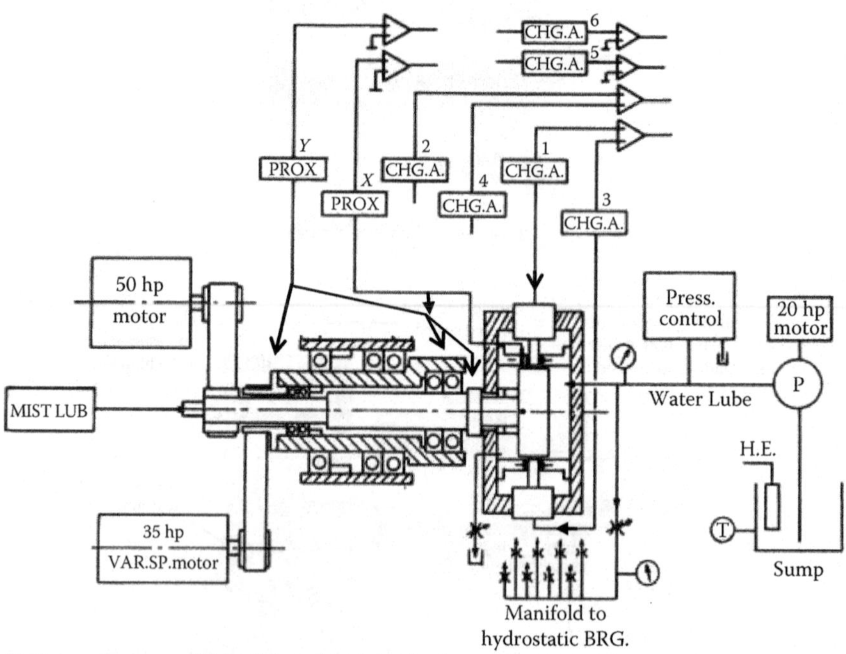

CHG.A. = Charge amplifier of piezoelectric load cell
PROX = Proximity displacement probe
MIST LUB = Oil-mist lubrication connection

FIGURE 1.4
Data acquisition for testing bearing or interstage fluid annuli.

to the fluid-annulus-pressurized test chamber. Also not shown is the 11,000 gallon closed-loop water sump tank, which is sized to be a large heat sink so that sump cooling was unnecessary over the duration of a test run. Water viscosity was set by heating the sump water to a specified temperature prior to the test run. The bearing/seal test chamber is designed to be hermetically sealed for testing seals with large through flows and pressure drops, or open to ambient as is typical for testing journal bearings. The bearing or seal test specimen can be either very stiffly held by piezoelectric load measuring cells or flexible mounted. A full description of the complete test facility and data processing steps are given by Adams et al. (1988, 1992) and Sawicki et al. (1997).

A close examination of Figure 1.2(c) reveals that the double-spool outer spindle is comprised of two closely fitted cylinders. These two cylinders are precisely ground to have an accurately *adjustable* centerline radial eccentricity range with respect to each other of 0–0.030 inch (0.76 mm). This eccentricity adjustment is manually set using two hand spanner wrenches to rotate one cylinder with respect to the other, with, of course, the spindles not spinning. As shown in Figures 1.2 and 1.4, each spindle spool has its own poly V-belt drive end for an independently controlled *variable speed* to each spindle spool. This feature permits the rotational centerline of the inner spool shaft to have a circular orbit frequency that is independently controlled, separate from the inner rotor's controlled variable speed of rotation. *Orbit-frequency-to-rotational-speed* ratio controllability is thus provided, and this yields a test-parameter matrix sufficiently diverse to extract anisotropic and nonsymmetric radial stiffness, damping, and virtual-mass linearized 2×2 rotor dynamic coefficient matrices (Equation 1.1) for the tested fluid annulus.

The instantaneous radial fluid dynamic force vector upon the nonrotating fluid annulus ring surface is measured with four *very stiff* (i.e., negligible radial deflection) piezoelectric force transducers in two mutually perpendicular x–y radial directions (at 45° and 135° relative to the horizontal floor), with both uniaxial pairs preloaded against each other to prevent their piezoelectric load cell crystals from ranging out of preload compression. The fixtures holding these four radial force transducers (Figure 1.5) are geometrically configured to be *very stiff* in the radial direction but *soft* in the axial and circumferential directions, that is, one piece consisting of two I-beams at 90° to each other as illustrated. That feature prevents any annular fluid-friction, torque-restraining force from contaminating the radial force signals. As illustrated, each of the four radial force fixtures are also configured with a hermetically sealed in-a-hole 4-strain-gage temperature-compensating bridge arrangement to measure the static radial load vector on the test ring. As illustrated in Figure 1.2(c), the adjustability of the inner spindle's circular-orbit radius magnitude (0–0.030 inch, 0.76 mm) facilitates setting the orbit radius *sufficiently large* to achieve an acceptable measured-force, signal-to-noise ratio, but *small enough* to minimize *dynamic nonlinearity* in the orbit-produced fluid-annulus dynamic force.

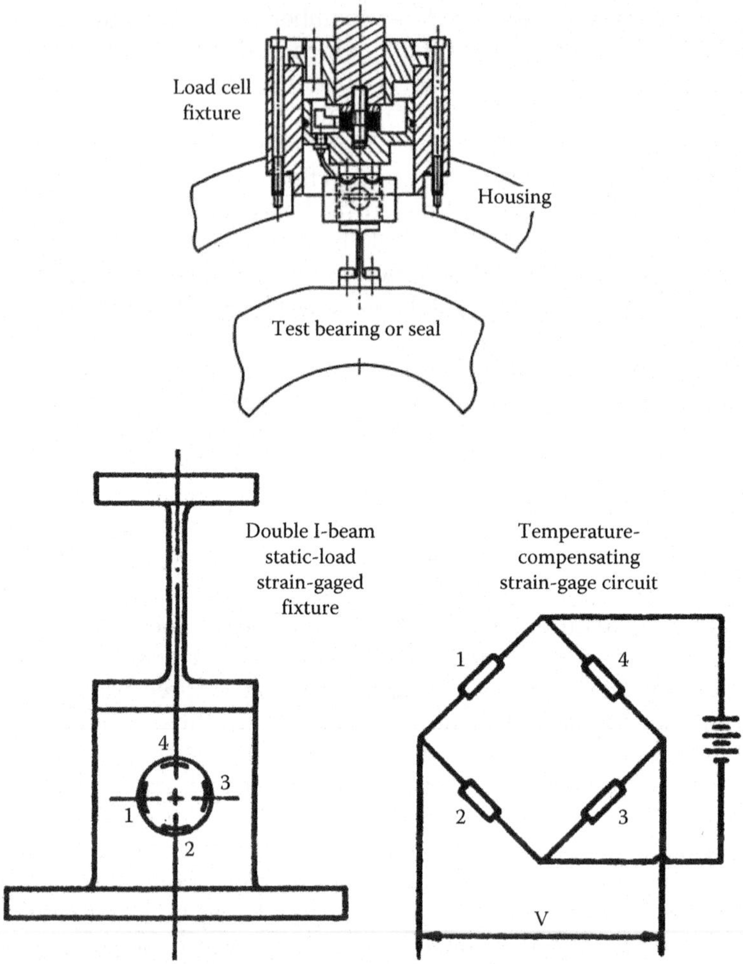

FIGURE 1.5
Dual piezoelectric/strain-gage radial load measuring fixture.

Other notable important features of this test apparatus include the precision hand-lapped-together, machine-tool-type tapered-fit attachment of the test journal to the inner spool shaft (Figure 1.2c), so that a test journal can be removed and subsequently reinstalled without incurring significant runout. For ultraprecision runout removal, the test journal is finish-ground with a back-and-forth axially oscillating grinder spindle while the internal spool shaft is slowly spinning at approximately 50 rpm. Any journal-targeting displacement *proximity-probe* indicated runout remaining after doing all of that is attributed to "electrical runout" and is postprocess removed digitally from the proximity probe raw test signals (Horattas et al. 1997).

Figure 1.4 illustrates the acquisition of measured force and displacement signals digitized and stored in real time. All these signal acquisitions are triggered by a 500-slotted optical-encoder interruption disk attached to the outboard end of the outer spool shaft. Thus, even with any slight outer-spool speed drift, the same number of digitized data points are captured for each cycle of rotor orbit. Postprocess *time averaging* is then easily used to filter out any signal noise and inner spindle spin speed harmonics that are not coherent with the outer spool's orbit-producing period of rotation.

1.1.1 Extracting Anisotropic Radial Rotor Dynamic Coefficient Matrices

The standard method by which rotor–stator interactive radial dynamic forces are modeled in rotor vibration analyses utilizes linearized spring stiffness $[K_{ij}]_{2\times2}$, damping $[C_{ij}]_{2\times2}$, and virtual-mass $[M_{ij}]_{2\times2}$ connecting-element coefficient matrices that are more rigorously covered elsewhere by the author (2010). For oil-film journal bearings characterized by the Reynolds Lubrication Equation (RLE), only stiffness and damping elements are typically employed because of the RLE's justified neglect of both *temporal* and *convective* fluid inertia effects. For higher-Reynolds-number fluid annuli, the additional linearized virtual-mass-connecting fluid inertia effects $[M_{ij}]_{2\times2}$ are included but constrained to symmetry (Adams 1987). This general linearized *rotor–motion* interaction force model is given in Equation 1.1. Using the generic journal bearing as an example, Figure 1.3 illustrates the applied static load vector $\vec{W}$ and the fluid-annulus pressure-distribution total reaction force $\vec{F}$ that includes both a static-load equilibrating force component $-\vec{W}$ plus the instantaneous time-varying rotor dynamic perturbation force $\vec{f} = \vec{W} + \vec{F}$. Therefore, relative to the static equilibrium position, $\vec{f}$ is formulated as follows:

$$\begin{Bmatrix} f_x \\ f_y \end{Bmatrix} = -\begin{bmatrix} k_{xx} & k_{xy} \\ k_{yx} & k_{xx} \end{bmatrix}\begin{Bmatrix} x \\ y \end{Bmatrix} - \begin{bmatrix} c_{xx} & c_{xy} \\ c_{yx} & c_{yy} \end{bmatrix}\begin{Bmatrix} \dot{x} \\ \dot{y} \end{Bmatrix} - \begin{bmatrix} m_{xx} & m_{xy} \\ m_{yx} & m_{yy} \end{bmatrix}\begin{Bmatrix} \ddot{x} \\ \ddot{y} \end{Bmatrix} \tag{1.1}$$

where $k_{ij} \equiv -\dfrac{\partial F_i}{\partial x_j}$, $c_{ij} \equiv -\dfrac{\partial F_i}{\partial \dot{x}_j}$, and $m_{ij} \equiv -\dfrac{\partial F_i}{\partial \ddot{x}_j}$.

In the fullest application of the apparatus shown in Figures 1.2 and 1.4, all four stiffness, four damping, and four inertia coefficients shown in Equation 1.1 can be *extracted* for a tested bearing or other fluid annulus. The *inputs* are the x and y radial harmonic displacement signals of the journal and the *outputs* are the x and y radial force signals required to rigidly position the bearing or seal relative to the rotating cylindrical piece. If the bearing or seal has orbital motion that can't be neglected, then the *inputs* are the x and y radial displacement signals of the journal relative to the bearing and the inertia effect of the test bearing or the seal mass (i.e., D'Alembert force) must be subtracted from the *output* measurements of the x and y load cell forces that support the test bearing or seal. Although the apparatus shown in

Figures 1.2 and 1.4 produces an orbit that is very close to perfectly circular, it is not required that the orbit be assumed perfectly circular because the orbit is precision-measured with the previously described multiprobe compliment of noncontacting proximity probes. The mechanical impedance model (Adams 1987, 2010) postulates that the measured x and y orbit displacement signals (inputs) and the resulting dynamic force signals (outputs) are all harmonic. The processed signals extracted from the measurements can thus be expressed as in Equation 1.2.

$$x = Xe^{i(\Omega t+\phi_x)}, \quad y = Ye^{i(\Omega t+\phi_y)}, \quad f_x = F_x e^{i(\Omega t+\theta_x)}, \quad f_y = F_y e^{i(\Omega t+\theta_y)} \tag{1.2}$$

Ω is the orbital frequency, $\mathrm{i} = \sqrt{-1}$, Ω and spin speed ω are independent.

Substituting Equation 1.2 into Equation 1.1 yields two complex equations. Using the basic formula $e^{iz} = \cos z + i \sin z$ separates the *real* and *imaginary* parts of those two complex equations, thereby yielding the following four real equations:

$$\begin{aligned}
F_x \cos\theta_x &= \left[\left(\Omega^2 m_{xx} - k_{xx}\right)\cos\phi_x + c_{xx}\Omega\sin\phi_x\right]X \\
&\quad + \left[\left(\Omega^2 m_{xy} - k_{xy}\right)\cos\phi_y + c_{xy}\Omega\sin\phi_y\right]Y \\
F_x \sin\theta_x &= \left[\left(\Omega^2 m_{xx} - k_{xx}\right)\sin\phi_x - c_{xx}\Omega\cos\phi_x\right]X \\
&\quad + \left[\left(\Omega^2 m_{xy} - k_{xy}\right)\sin\phi_y - c_{xy}\Omega\cos\phi_y\right]Y \\
F_y \cos\theta_y &= \left[\left(\Omega^2 m_{yx} - k_{yx}\right)\cos\phi_x + c_{yx}\Omega\sin\phi_x\right]X \\
&\quad + \left[\left(\Omega^2 m_{yy} - k_{yy}\right)\cos\phi_y + c_{yy}\Omega\sin\phi_y\right]Y \\
F_y \sin\theta_y &= \left[\left(\Omega^2 m_{yx} - k_{yx}\right)\sin\phi_x - c_{yx}\Omega\cos\phi_x\right]X \\
&\quad + \left[\left(\Omega^2 m_{yy} - k_{yy}\right)\sin\phi_y - c_{yy}\Omega\cos\phi_y\right]Y
\end{aligned} \tag{1.3}$$

Since there are 12 *unknowns* in these four equations (i.e., four stiffness, four damping, and four inertia coefficients), measured data must be obtained at a minimum of three *discrete orbit frequencies* for a given equilibrium operating condition. There are a number of data reduction ("curve fitting") approaches when test data is taken at a multitude of orbit frequencies for a given equilibrium operating condition. For example, a frequency-localized 3-frequency fit propagated over a frequency range with several frequency data points produces *frequency-dependent* stiffness, damping, and inertia coefficients to the extent that it improves the fitting of the measurements to the impedance model in Equation 1.3. However, it is more end-user-friendly to reduce the measurement data using a *least squares linear regression* fit of all the measured data over the full tested frequency range, because that yields all 12 extracted coefficients as "constants" independent of frequency, which is far less cumbersome as inputs for rotor vibration analysis computer codes.

By its proven performance and excellent repeatability, the apparatus shown in Figures 1.2 and 1.4 is very accurate and very close to "linear." Data is collected at 50–100 consecutive cycles of orbit excitation frequency Ω and then *time averaged* to remove all noise and other noncoherent signal content, that is, spin-speed ω mechanical harmonics. *Time averaging* is simply summing up each of the signal's magnitudes for N consecutive Ω-frequency cycles and then dividing each of these sums by the integer N. The time-averaged signals from each measurement channel are then Fourier series decomposed. When Ω-frequency components are much larger than all the n Ω harmonics, it indicates a high degree of linearity of the apparatus and test condition.

1.1.2 Hybrid Hydrostatic–Hydrodynamic Bearing Dynamic Properties

Test results on a hybrid hydrodynamic–hydrostatic journal bearing by Adams et al. (1992) are in close agreement with theoretical predictions by Childs (1992) and Sawicki et al. (1997). An illustration of the tested bearing is shown in Figure 1.6.

Samples of time-averaged measured bearing force signals are shown in Figure 1.7. The data sample in Figure 1.7(b) for one cycle shows only a quite small higher harmonic content of the orbital frequency of the journal, thus

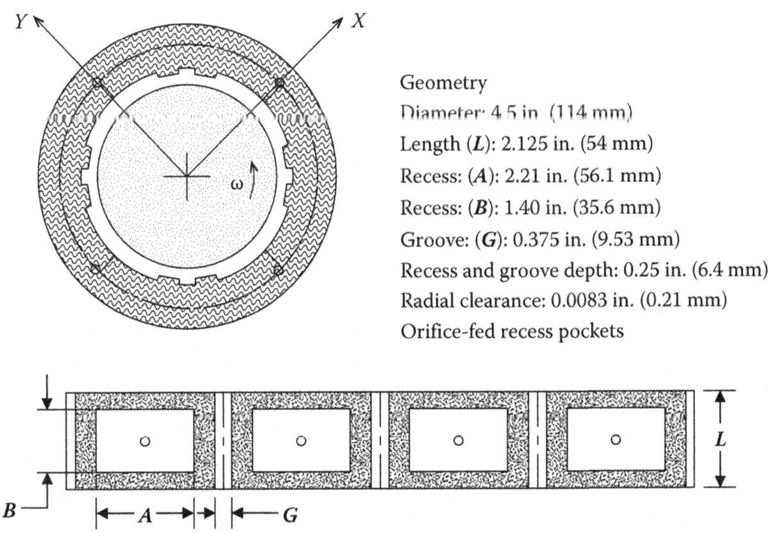

FIGURE 1.6
Hybrid hydrostatic–hydrodynamic test journal bearing.

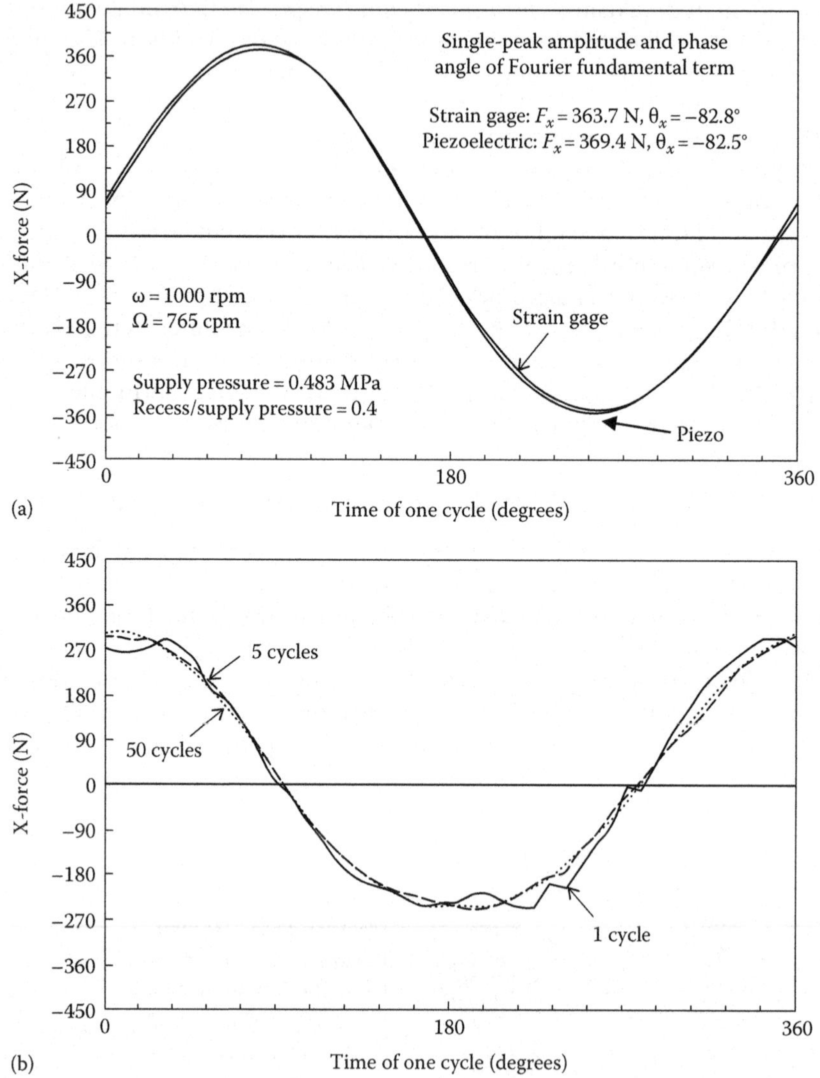

FIGURE 1.7
(a) Comparison between strain-gage and piezoelectric load cell measurements of the same bearing radial force and (b) time averaging of measured bearing dynamic force cycles (note that 5 cycles and 50 cycles of time-averaged force signals are indistinguishable).

demonstrating a high degree of linearity in the entire test setup. Furthermore, the data sample in Figure 1.7(b) shows that by time averaging 50 consecutive cycles of data, the higher harmonics were filtered out prior to inputting their F_x and F_y peak amplitudes and θ_x and θ_y phase angles into Equation 1.3. The sample comparison in Figure 1.7(a) shows quite good agreement between the two independent time-averaged bearing force measuring sensors: strain-gage and piezoelectric. Figure 1.7(b) shows the rapid convergence for the time averaging of consecutive cycles.

1.1.3 Journal-on-Bearing Impact Restitution Coefficient

One ingredient for computational simulations of nonlinear rotor vibration phenomena includes the modeling of impacts between rotating and nonrotating components, for example, between journal and bearing. Impact between two solid bodies has long been known to be a quite complex phenomenon at the fundamental deformation level. Modern Finite Element elastic-plastic models now provide detailed simulations of impact between complex deformable bodies, for example, auto collisions and artillery penetration of armaments. But more often, modeling of impact forces is still performed using the traditional empirically based approach. In that approach, the major input is the well-known empirical *coefficient of restitution*, which computationally reduces the modeling of the inherent complexity of impact dynamics to a simple analysis formulation.

The test apparatus described in Figure 1.2 was augmented with additional components and instrumentation to experimentally determine the *coefficient of restitution* for impacts between a journal bearing and a journal. The full description of this research is given by Adams et al. (2000). The augmented test apparatus is illustrated in Figure 1.8. The two biggest challenges in this endeavor were (1) how to accurately measure X and Y orbital velocity signals through a rotor–stator impact event and (2) how to experimentally produce a

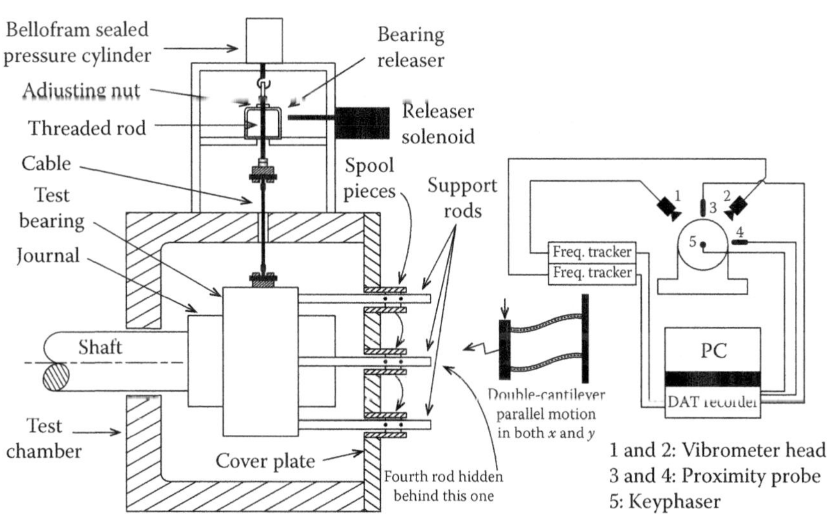

FIGURE 1.8
Impact test for rotor–stator restitution coefficient measurements.

true line contact (and alternatively a specified axial misalignment edge contact) between the bearing sleeve and journal at the moment of impact.

To accurately measure bearing orbital velocity directly through the very short time duration of a rotor–stator impact, two orthogonal channels (X and Y) of laser fiber optical vibrometers were employed (Figure 1.9). At the same time, two orthogonal channels of noncontacting-inductance-type proximity probes were employed to measure bearing displacement (Figure 1.10). Using test data from the qualification test shown in Figure 1.9, a wavelet-transform-based high-pass filtering of the first numerical time derivative of the digital displacement signal was compared to the direct-laser-measured velocity.

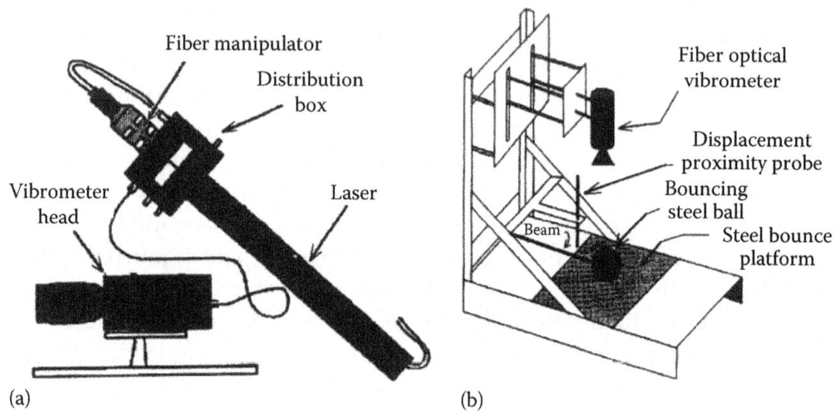

FIGURE 1.9
Laser vibrometer qualification testing: (a) Laser vibrometer system components and (b) qualification test for laser vibrometer.

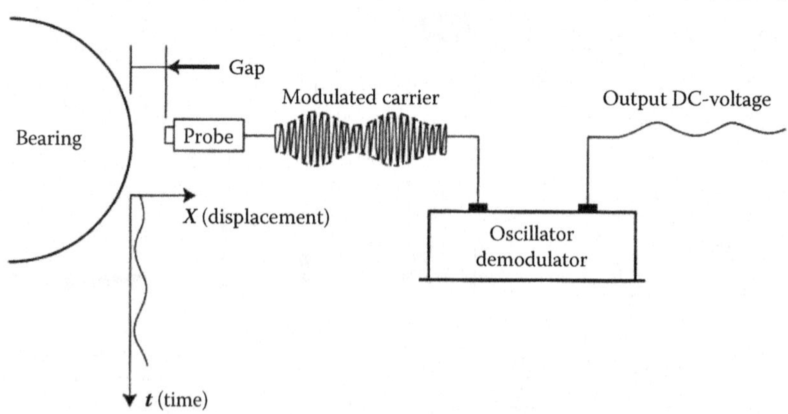

FIGURE 1.10
Inductance eddy current noncontacting position-sensing system.

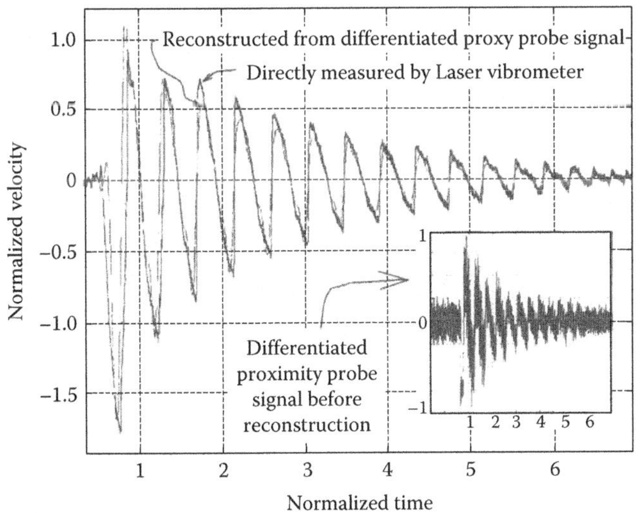

FIGURE 1.11
Comparison between the impact velocity signal from a direct laser vibrometer signal and a reconstructed differentiated displacement signal.

A sample of this qualification testing is shown in Figure 1.11, validating the accurate functioning of the two laser vibrometers.

The fixtures used to experimentally produce a true line contact (and alternatively a specified axial misalignment edge contact) between stator sleeve and journal at the moment of impact are illustrated in Figure 1.8. Four double cantilever beams provide the parallel motion of the bearing in the radial plane. So for true line contact, while the bearing is sat freely upon the journal, the eight set screws for each of the four support rods (double cantilever beams) are carefully engaged into their respective support rod, with two high-precision dial indicators used to insure that the control rods are not moved in the process. For a specified axial misalignment edge contact test case, the bearing is also sat freely upon the journal but with an intervening feeler gage to provide the desired misalignment. The support-rod-set screws are likewise engaged to provide parallel bearing displacement in the radial plane while maintaining the locked-in prescribed misalignment.

The bearing is released by a solenoid energized rod, at which point it drops upon the stationary journal. Prior to releasing the bearing to drop, a Bellowfram-sealed pressurized cylinder lifts up the bearing an amount to yield the desired initial impact velocity. A sample of the bearing and journal motion signals immediately before, during, and immediately after an impact test is shown in Figure 1.12. The crucial data is in the indicated "impact zone." The restitution coefficient is expressible as the ratio of the relative velocity of

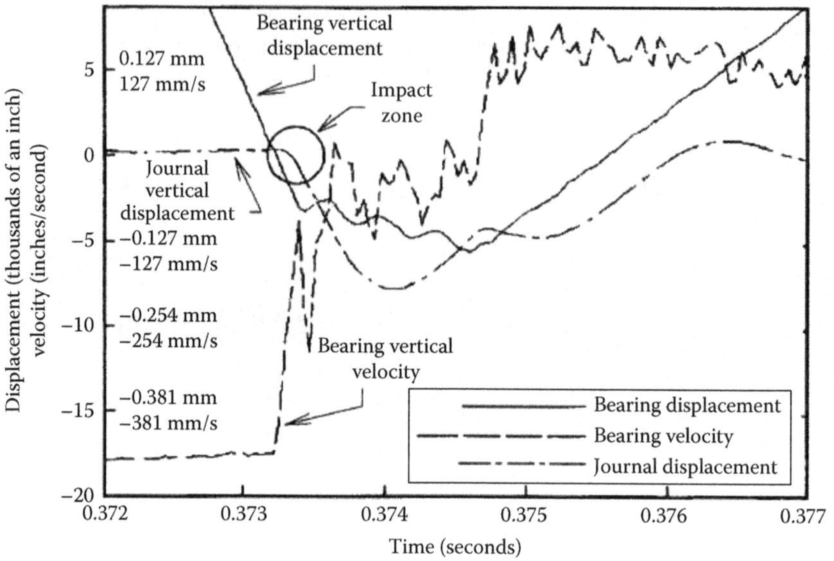

FIGURE 1.12
Bearing and journal motion before, during, and after impact.

the two masses after impact to their relative velocity before impact as in the following equation:

$$e = \frac{V_B' - V_R'}{V_B - V_R} \tag{1.4}$$

where

V_B and V_R are the bearing and rotor velocities, respectively, just before impact
V_B' and V_R' are the bearing and rotor velocities, respectively, just after impact

The test results of line-contact impact are given in Figure 1.13 and show restitution coefficient values typical for impact tests (Goldsmith 1960). Prior to these tests, a project was completed to compare impacting rotor motions on a small flexible-rotor test rig with time-transient nonlinear simulations for that rig. A restitution coefficient of 0.8 was assumed for those simulations. In Section 2.2 and also given by Adams et al. (2000), the comparisons between those tests and simulations show the simulations having somewhat more "bounce" than the corresponding test cases. This is consistent with the test results in Figure 1.12. That is the restitution coefficient results in Figure 1.12 are consistently lower than 0.8 and suggest that nonlinear simulations should accommodate a variable restitution coefficient that is a function of the instantaneous rotor-to-stator impact velocities.

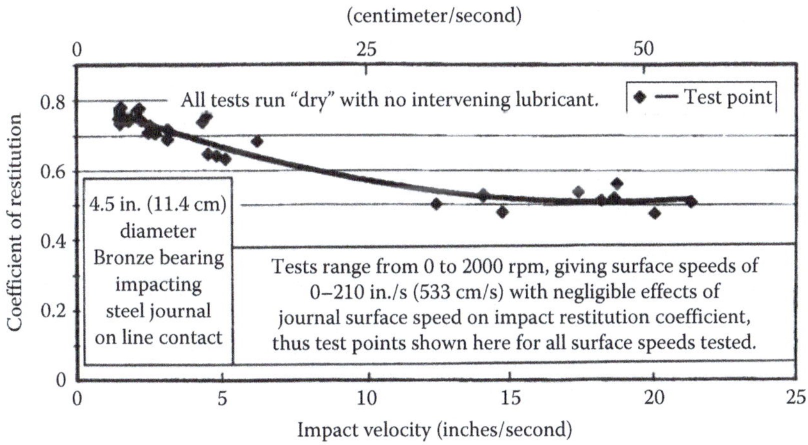

FIGURE 1.13
Bearing-on-journal impact restitution coefficient; bearing radial clearance = 0.010 inch (0. 254 mm).

1.2 Double-Spool Spindle Rig at Caltech for Pump Stage Testing

The spindle assembly shown in Figure 1.14 was developed by researchers at the California Institute of Technology exclusively for a NASA-sponsored research focused primarily on the Space Shuttle's very-high-speed centrifugal pumps, such as those for main-engine liquid hydrogen. Of course, this test rig used water, not liquid hydrogen, as the working fluid. An impeller rotational speed of up to 35,000 rpm was possible and orbit frequencies up to 60 Hz. As delineated by Adams (2010), the fluid dynamic forces that naturally act upon centrifugal pump impellers are comprised of three distinguishable types: (1) static, (2) dynamic unsteady-flow-induced time-varying, and (3) dynamic in response to impeller vibratory motion.

1.2.1 Rotor Dynamic Properties of a Complete Centrifugal Pump Stage

The Caltech research in this area was instigated primarily to investigate the Space Shuttle hydrogen pump high-amplitude self-excited rotor vibration discovered during the development testing of these pumps for NASA. The details of this research are reported by Chamieh et al. (1982, 1984). Adams (2010) focuses on the rotor dynamic results of this Caltech research, in particular the linearized radial stiffness, damping, and virtual mass (fluid inertia) coefficients, k_{ij}, c_{ij} and, m_{ij}, respectively, as shown in Equation 1.1 for the

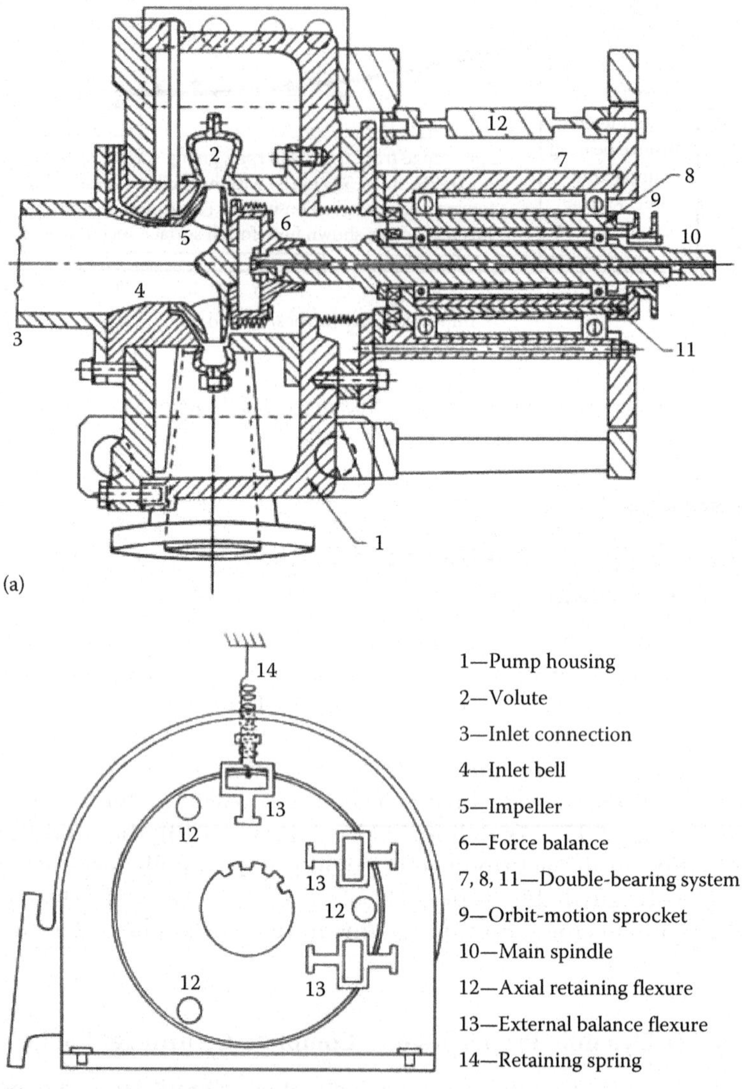

FIGURE 1.14
Cross section of a Caltech double-spool spindle for centrifugal pump impeller fluid dynamical force measurements: (a) plan view and (b) axial view.

TABLE 1.1

Impeller Rotor Dynamic Radial Force Dimensionless Coefficients

Source/Type	$\bar{k}^s$	$\bar{k}^{ss}$	$\bar{c}^s$	$\bar{c}^{ss}$	$\bar{m}^s$	$\bar{m}^{ss}$
Caltech/volute	−2.5	1.1	3.14	7.91	6.51	−0.58
Caltech/diffuser	−2.65	1.04	3.80	8.96	6.60	−0.90
Sulzer/diffuser (2000 rpm)	−5.0	4.4	4.2	17.0	12.0	3.5
Sulzer/diffuser (4000 rpm)	−2.0	7.5	4.2	8.5	7.5	2.0

$\bar{k}_{ij} \equiv \dfrac{k_{ij}}{\pi\rho R_2 B_2 \omega^2}$, $\bar{c}_{ij} \equiv \dfrac{c_{ij}}{\pi\rho R_2 B_2 \omega}$, $\bar{m}_{ij} \equiv \dfrac{m_{ij}}{\pi\rho R_2 B_2}$

k_{ij}, Dimensional stiffness; c_{ij}, dimensional damping; m_{ij}, dimensional inertia; ρ, mass density of pumped liquid; R_2, impeller outer (discharge) radius; B_2, impeller discharge width including impeller side plates; ω, impeller spin speed; *s*, symmetric; *ss*, skew-symmetric.

general *anisotropic* model, and for the *isotropic* rotor vibration model in the following equation:

$$\begin{Bmatrix} f_x \\ f_y \end{Bmatrix} = -\begin{bmatrix} k^s & k^{ss} \\ -k^{ss} & k^s \end{bmatrix}\begin{Bmatrix} x \\ y \end{Bmatrix} - \begin{bmatrix} c^s & c^{ss} \\ -c^{ss} & c^s \end{bmatrix}\begin{Bmatrix} \dot{x} \\ \dot{y} \end{Bmatrix} - \begin{bmatrix} m^s & m^{ss} \\ -m^{ss} & m^s \end{bmatrix}\begin{Bmatrix} \ddot{x} \\ \ddot{y} \end{Bmatrix} \quad (1.5)$$

where superscript *s*—symmetric and *ss*—skew-symmetric.

In Adams (2010), a compelling fundamentals-based argument is made to support the stipulation that in reducing experimental force and corresponding motion test data to extract the isotropic coefficients in Equation 1.5, $m^{ss} = 0$ should be imposed. The comparison, Table 1.1, between Caltech and Sulzer Pump Co. Guelich et al. (1993), similarly extracted experimental results, when $m^{ss} = 0$ is not imposed, supports Adams' $m^{ss} = 0$ postulate. Specifically, the comparisons are consistent and compare reasonably for both sources, except for the glaring inconsistency in the m^{ss} coefficients, that is, the Caltech results all have $m^{ss} < 0$, while the Sulzer results all have $m^{ss} > 0$.

1.2.2 Unsteady-Flow Forces

Unsteady-flow forces are always present in centrifugal pumps. At operating pump flows away from the best efficiency point (BEP), the unsteady flow forces are much larger than at BEP operation, particularly in high-energy-density centrifugal pumps like power plant feed water pumps (as large as 80,000 hp) (Makay et al. 1978, 1980, 1984) and the Space Shuttle's main engine pumps. The "snapshot" illustrations of impeller flow paths in Figure 1.15 provide a visual appreciation of high-energy-flow-induced unsteady-flow dynamic forces. These dynamic forces are capable of destroying pump internals. So quantifying such unsteady-flow forces and their response to design parameter options was already widely recognized by the 1970s. One of

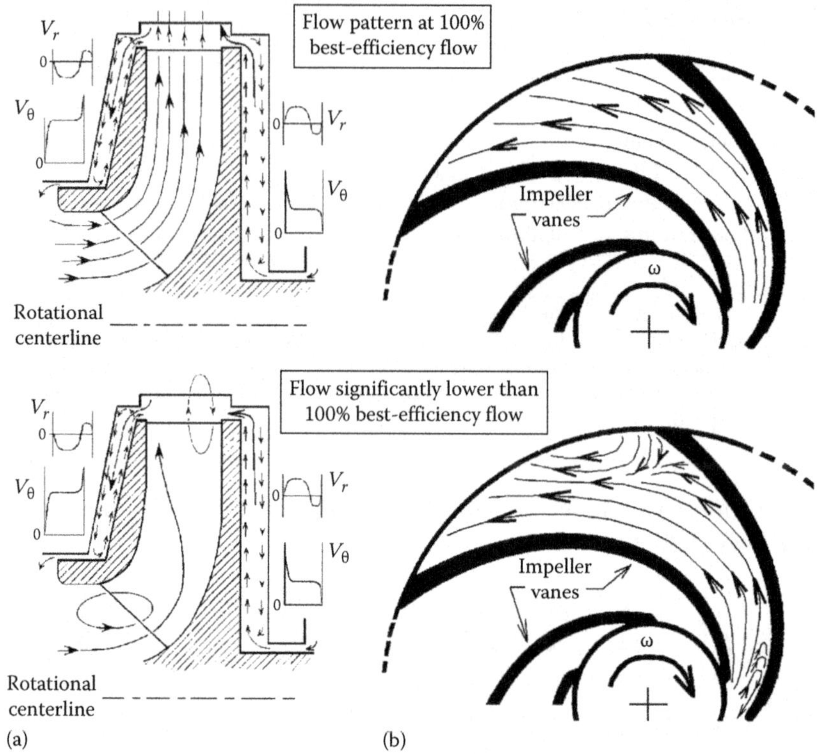

FIGURE 1.15
Snapshots of centrifugal pump impeller flow patterns. (a) Radial flow views and (b) circumferential flow views relative to rotating impeller.

NASA's objectives for the Caltech dual-spindle test facility was specifically to significantly extend knowledge in this area of centrifugal pump technology. Likewise, this was also one of the major objectives of the EPRI-sponsored \$10M feed pump research project with the Sulzer Pump Company (Guelich et al. 1993). A detailed coverage of this research and results are given by Adams (2017).

1.3 Axial Bearing Squeeze-Film Damper Concept and Development

As clearly explained by Dr. Elemer Makay in his many short courses and publications (e.g., Makay et al. 1978, 1980, 1984), those same unsteady-flow phenomena that cause high-magnitude radial impeller forces, particularly at low-flow operation, also cause high axial dynamic impeller forces that in turn result in high levels of axial vibration. This prompted an EPRI-sponsored task to explore experimental approaches to thrust-bearing squeeze-film dampers similar to the radial-bearing squeeze-film dampers commonly employed in modern aircraft gas turbine jet engines. The major product of that exploratory research (Adams 1983) is shown in Figure 1.16.

This rig employs a standard off-the-shelf, double-acting, tilting-pad, Kingsbury-type thrust bearing of typical feed pump size, supported in an annular axial squeeze-film damper in parallel with Belville springs. This is comparable with a radial bearing squeeze-film damper employing centering springs. As shown, the thrust collar is rigidly attached to a rotating shaft, which is supported by two hydrostatic journal bearings with very high radial stiffness, but allowing the test shaft to move freely in the axial direction. The end of the test shaft is coupled, through very stiff rolling contact bearings, to the nonrotating output ram of a high-force capacity hydraulic actuator. The hydraulic actuator is a standard off-the-shelf product, sized and controlled so that the tested thrust bearing damper configuration can be subjected to the combinations of static and dynamic loading that occurs in actual large feed water pumps. The objective here was to develop and prove a new design concept in feed pump thrust bearing support that would significantly attenuate axial rotor vibration and transmitted axial dynamic forces. Meeting this objective was thought to significantly reduce or virtually eliminate pump failures and accelerated wear due to axially transmitted unsteady-flow dynamic forces. Due to funding limitations, this test apparatus was not built.

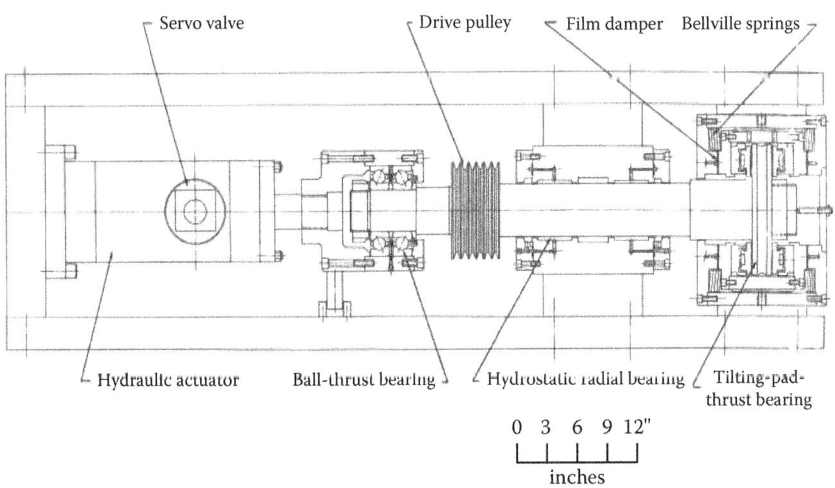

FIGURE 1.16
Apparatus design of a thrust-bearing squeeze-film damper test.

Bibliography

Adams, M. L., Development of advanced rotor-bearing systems for feedwater pumps—Phase III: Hardware design and fabrication, EPRI Final Report CS-3203, EPRI Project 1884-4, Electric Power Research Institute, Palo Alto, CA, 45pp., July 1983.

Adams, M. L., Insights into linearized rotor dynamics, Part-2, *Journal of Sound & Vibration*, 112(1), 97–110, 1987.

Adams, M. L., *Rotating Machinery Vibration: From Analysis to Troubleshooting*, 2nd edn., Taylor & Francis, CRC Press, Boca Raton, FL, 465pp., 2010.

Adams, M. L., *Power Plant Centrifugal Pumps: Problem Analysis and Trouble-Shooting*, Taylor & Francis, CRC Press, Boca Raton, FL, 182pp., 2017.

Adams, M. L., Afshari, F., and Adams, M. L., An experiment to measure the restitution coefficient for rotor-stator impacts, *Seventh IMechE International Conference on Vibration in Rotating Machinery*, Nottingham, England, pp. 301–308, Sept. 2000.

Adams, M. L., Sawicki, J. T., and Capaldi, R. J., Experimental determination of hydrostatic journal bearing rotordynamic coefficients, *Proceedings, Fifth IMechE International Conference on Vibration in Rotating Machinery*, Bath, England, pp. 365–374, Sept. 1992.

Adams, M. L., Yang, T., and Pace, S. E., A sea rest facility for the measurement of isotropic and anisotropic linear rotordynamic characteristics, *Proceedings of NASA Sponsored Workshop on Rotordynamic Instability Problems in High-Performance Turbomachinery*, Texas A&M University, College Station, TX, NASA CP-3026, May 1988.

Chamieh, D., Acosta, A. J., and Caughey, T. K., Experimental measurements of hydrodynamic stiffness matrices for a centrifugal pump impeller, *Workshop: Rotordynamic Instability Problems in High Performance Turbomachinery*, Texas A&M University, College Station, TX, NASA CP No. 2250, 1984.

Chamieh, D., Caughey, T. K., Brennen, C. E., and Acosta, A. J., Comments on impeller-volute interactions, *Proceedings, Power Plant Feed Pumps-State of the Art, EPRI Symposium*, Cherry Hill, NJ, June 1982.

Childs, D., Private communications with M. L. Adams at *Fifth IMechE International Conference on Vibration in Rotating Machinery*, Bath, England, Sept. 1992.

Goldsmith, G., *Impact—The Theory and Physical Behavior of Colliding Solids*, Edward Arnold Publishers, Ltd., London, U.K., 1960.

Guelich, J. F., Bolleter, U., and Simon, A., Feedpump operation and design guidelines, EPRI Final Summary Report TR-102102, Research Project 1884-10, Electric Power Research Institute, Palo Alto, CA, 1993.

Horattas, G. A., Adams, M. L., and Dimoftte, F., Mechanical and electrical runout removal on a precision rotor-vibration research spindle, *ASME Journal of Acoustics and Vibration*, 119(2), 216–220, 1997.

Makay, E., Survey of feed pump outages, EPRI Research Project RP 641, Final Report FP-754 (Edited by M. L. Adams), Electric Power Research Institute, Palo Alto, CA, p. 98, 1978.

Makay, E., How close are your feed pumps to instability-cased disaster, *Power Magazine*, pp. 69–71, Dec. 1980.

Makay, E. and Barrett, J., Changes in hydraulic component geometries greatly increased power plant availability and reduced maintenance costs: Case histories, *Proceedings, Texas A&M First International Pump Symposium*, College Station, TX, 1984.

Sawicki, J., Capaldi, R. J., and Adams, M. L., Experimental and theoretical rotordynamic characteristics of a hybrid journal bearing, *ASME Journal of Tribology*, 119(1), 132–142, Jan. 1997.

2

Tabletop Rigs: Bently-Nevada Rotor Kit and Automatic 2-Plane Rotor Balancing

Bently-Nevada (*B-N*), part of General Electric since 2002, has been a major developer and producer of machinery vibration–monitoring sensors, instrumentation, data acquisition, and signal analysis products during the emergence of vibration-measuring systems in the second half of the twentieth century. Their products are used extensively in monitoring machinery in power plants, chemical process machinery, and manufacturing systems just to name a few. *B-N* was founded in 1961 by the late Dr. Donald E. Bently (1924–2012), who built up his company and closely managed it until its acquisition by GE in 2002. Throughout his pioneering work, Don Bently maintained long-standing interactions with rotor dynamics technologists worldwide, especially academics, like this author. It was one of the most exciting professional acquaintances of this author's career—to closely associate with my friend Don Bently. I'm surely not alone in this.

Don designed a small tabletop two-bearing rotor kit (Figure 2.1) on which to demonstrate his eddy current induction noncontacting proximity probes (Figure 1.10) for accurately measuring rotor vibration. Although he marketed this little test rig throughout the industry, he donated them to rotor vibration researchers working in academia, such as this author and many others. The *B-N* rotor kit evolved in its own right, becoming quite amenable to modifications and enhancements to facilitate its use in serious rotating machinery vibration research. This chapter presents examples of this from the work of the author and his graduate students spanning several years.

2.1 Squeeze-Film Rotor Vibration Dampers

The first of these examples is reported by Quinn (1984). The *B-N* rotor kit was modified by replacing the two original dry bronze bushing radial bearings with small preloaded angular-contact ball bearings, each supported within a squeeze-film damper, small journal bearing-sized annular oil-filled clearance. Figure 2.2 illustrates the modifications made to the original *B-N* rotor kit. More detailed information on various squeeze-film damper configurations, with and without centering springs, is given in Adams (2010).

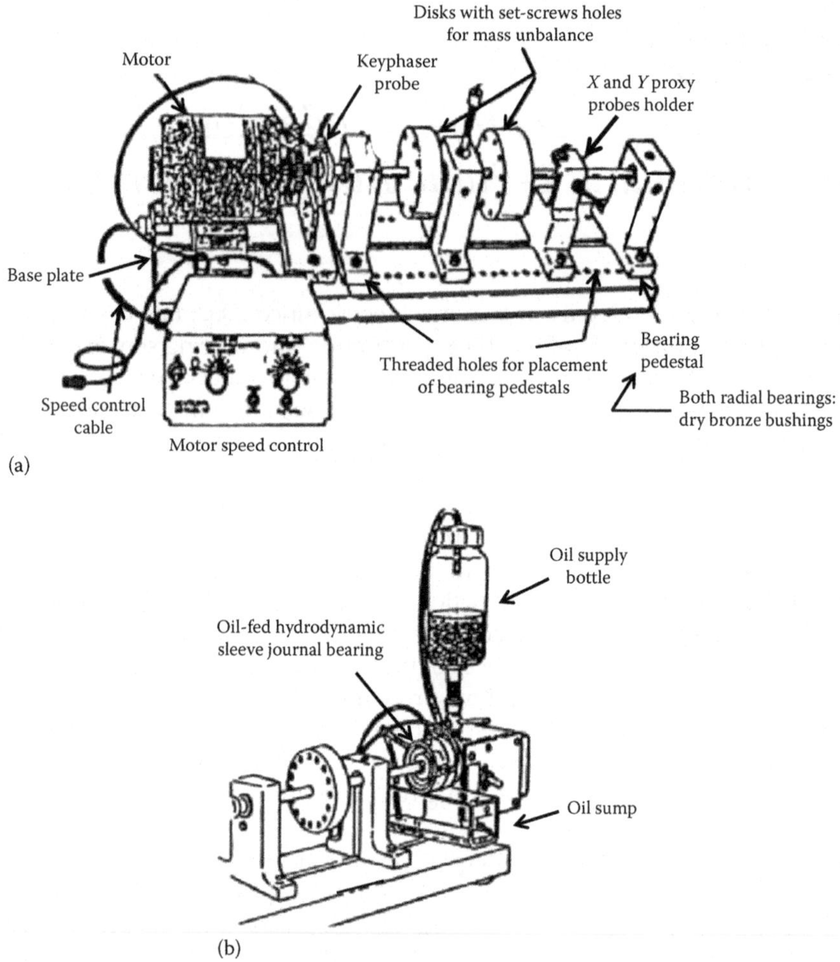

FIGURE 2.1
Bently-Nevada rotor kit: (a) original configuration and (b) journal bearing attachment to study self-excited, oil whip rotor vibration.

By adjusting the inlet oil supply pressure, tests were performed with varying degrees of cavitation within the damper film: from no cavitation to a mostly cavitated oil film. As covered by Adams (2010), the higher the frequency of rotor vibration, the less the cavitated oil-film region is able to condense back into the liquid state in synchronization with the vibration. That produces a cyclic hysteretic effect that causes a damper film liftoff away from the bearing, thus eliminating the need for centering springs. Therefore, it is common in many squeeze-film damper applications, such as modern aircraft gas turbine jet engines, not to use centering springs. As confirmed by test results from this modified *B-N* rotor kit

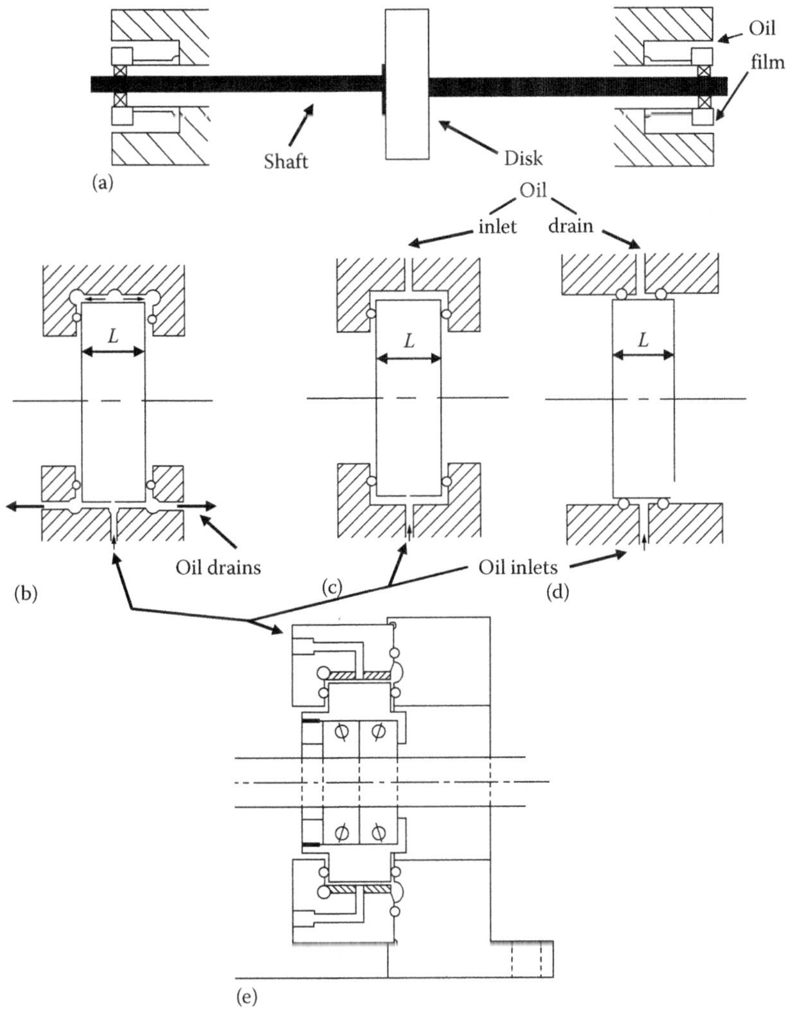

FIGURE 2.2
Modifications to the *B-N* rotor kit for squeeze-film damper research: (a) schematic of rotor-bearing configuration, (b) damper without a centering spring or pressurized axial boundary, (c) damper without a centering spring but with a pressurized axial boundary, (d) damper with a centering spring via O-rings and pressurized axial boundary, and (e) detailed configuration of a damper with a bearing.

in Figure 2.3, as the vibration magnitude is increased, the hysteretic liftoff effect becomes more pronounced.

A very important additional effect of cavitation in the damper film is demonstrated by the test results shown in Figure 2.4. Naturally, as the oil supply pressure is increased, the degree of oil film cavitation is lessened. With less oil film cavitation, the damper film becomes stiffer, and thus the rotor vibration

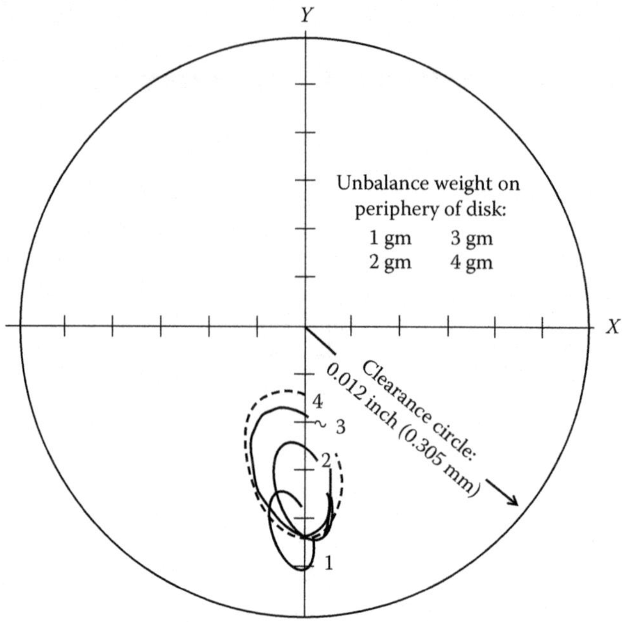

FIGURE 2.3
Modified *B-N* rotor kit vibration orbits at bearing damper location: rotor speed 9000 rpm; oil supplied at 80°F, 2 psig (1.4 N/cm^2 gage).

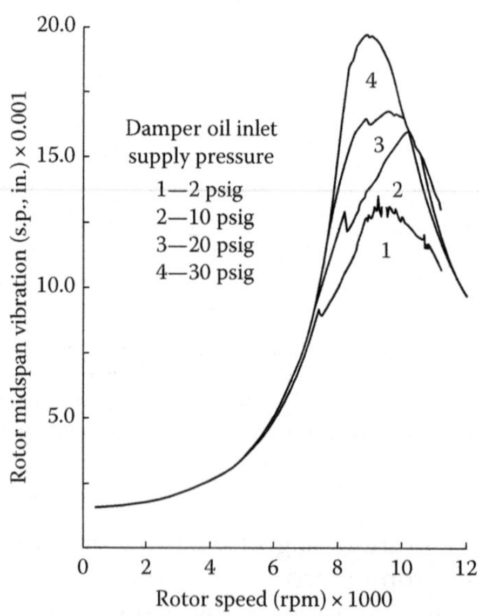

FIGURE 2.4
Midspan rotor radial vibration amplitude passing through the first critical speed at a progression of increases in damper oil inlet supply pressure.

amplitude at the damper becomes less. But that naturally lessens the amount of cyclic vibration energy dissipated by the damper. Consequently, with rotor flexibility, the vibration amplitude elsewhere on the rotor can be expected to possibly increase at resonance (critical speed). Therefore, the lessening of damper cavitation by increasing oil supply pressure should generally be expected to detrimentally reduce a damper's ability to quell rotor vibration through critical speeds.

2.2 Journal Bearing, Rub Impact, and Coefficient of Restitution

A long, slender, dynamically flexible shaft (second-generation *B-N* rotor kit) was the starting configuration. It was modified as illustrated in Figure 2.5(a)

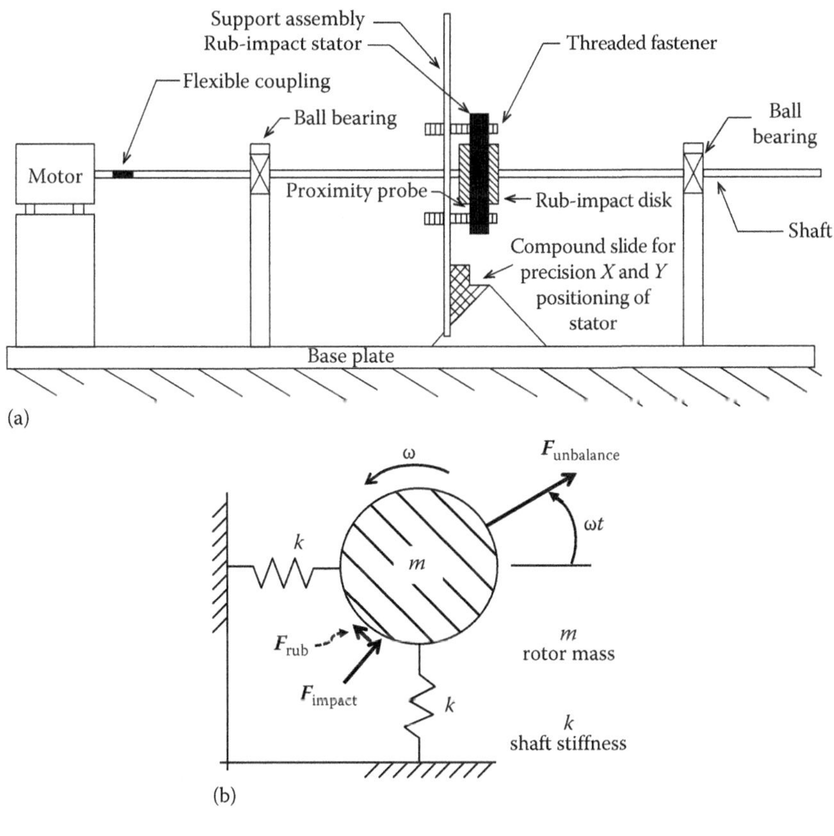

FIGURE 2.5
(a) Rub impact-modified *B-N* rotor kit and (b) simulation model.

(Adams 1996, Adams et al. 2000). A midspan rotor disk was employed as the rotating rub-impact component. The close clearance, nonrotating rub-impact component was held and radially positioned by a high-precision ***X*** and ***Y*** machine-tool compound slide as illustrated. The simulation model for computer-predicted nonlinear time-transient response is shown in Figure 2.5(b).

Figure 2.6 shows a sample of the results from this modified *B-N* rotor kit comparing computer simulations with the test results. Note that the simulations are a bit more "bouncy" than the test results. Referring back to Section 1.1.3 and Figure 1.13, restitution coefficient experimental results, the test results here in Figure 2.6 were obtained considerably prior to those shown in Figure 1.13. The restitution coefficient used to simulate the *B-N* rotor kit results in Figure 2.6 assumed a value of 0.8 for the restitution coefficient. This explains why the simulations in Figure 2.6 are more "bouncy" than the corresponding test results. Obviously, the assumed coefficient value of 0.8 was too high, as clearly shown by the much later obtained research results in Figure 1.13. This leads to an important conclusion. Specifically, when utilizing the Figure 1.13 experimental impact results for computational simulations of time-transient motions involving rotor–stator impacting, each impact force event should be computed utilizing a restitution coefficient value consistent with the instantaneous impact velocities.

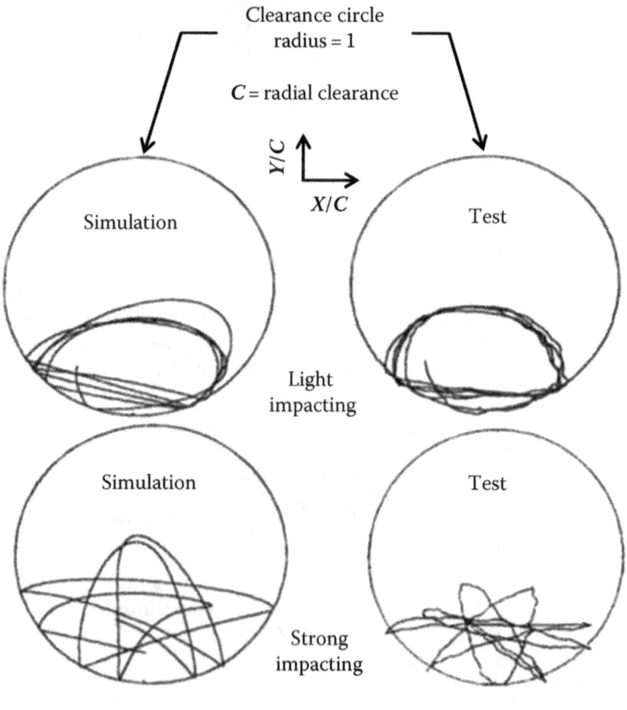

FIGURE 2.6
Comparisons between test and computer-generated dynamic simulations.

2.3 Nonlinear Hysteresis Loop of a Journal Bearing Oil Whip

The hysteresis loop associated with the journal bearing dynamic instability self-excited vibration phenomenon called *oil whip* was, for a long time, an interesting topic for the academics but did not attract the close scrutiny of rotating machinery development engineers. However, in the seismically active region of Japan, a team headed by Professor Y. Hori at the University of Tokyo brought the practical importance of the journal bearing hysteresis loop to the wider engineering community. Hori and Kato (1990) explain the distinct possibility of an *earthquake-initiated*, high-amplitude sustained, self-excited subsynchronous rotor vibration occurring in main steam turbines of large power plants. Their work inspired subsequent research by the author and his team, reported by Adams et al. (1996) and Horattas (1996).

A *B-N* rotor kit as illustrated in Figure 2.1 was employed for the experimental phase of this hysteresis loop research. As shown in Figure 1.2(b), the kit is supplied with a hydrodynamic journal bearing attachment to study self-excited oil whip rotor vibration. Figure 2.7(a) illustrates the modifications to the *B-N* rotor kit for this research. An upward, precisely controlled static load was maintained constant through a finely adjustable soft spring preload unit, Figure 2.7(b). Since that preload spring was relatively quite flexible, the upward static load that it applied to the shaft (to upwardly load the test journal bearing) did not significantly vary in response to rotor radial motion. The equivalent simulation model in Figure 2.7(c) was treated as a point mass rotor on two massless ***X*** and ***Y*** orthogonal radial support springs to model the thin flexible shaft's isotropic radial beam stiffness. Added to this was the time-transient, nonlinear, hydrodynamic journal bearing oil film force imbedded into a forward-marching time-step numerical integration of the rotor dynamic response, for example, as described by Adams (1980) and Adams and McCloskey (1984).

Figure 2.8 shows the journal bearing hysteresis loop for a substantial impose-bearing static radial load. As shown, it imbeds the classical small-perturbation-initiated oil whip phenomenon within the totality of possibilities. First, the small perturbation threshold speed ω_{th} is what mathematicians call a Hopf bifurcation point. Second, an expanded view shows two stable hysteresis loop vibration solutions (*zero* and *nonlinear limit cycle*) at speeds below the linear system classical oil whip threshold speed ω_{th} and one mathematically unstable solution that is bounded by the two mathematically stable solutions.

Adams et al. (1996) show simulation results for a wide range of nondimensional bearing static load. Their results show, as well known, that increasing bearing static load raises the expected oil whip threshold speed ω_{th}. But as not widely realized, increasing bearing static load also lowers the rotor speed above which a large-amplitude motion disturbance, such as an earthquake, can cause a large-amplitude oil whip limit-cycle sustained vibration. The example hysteresis loop in Figure 2.8 captures all this insight.

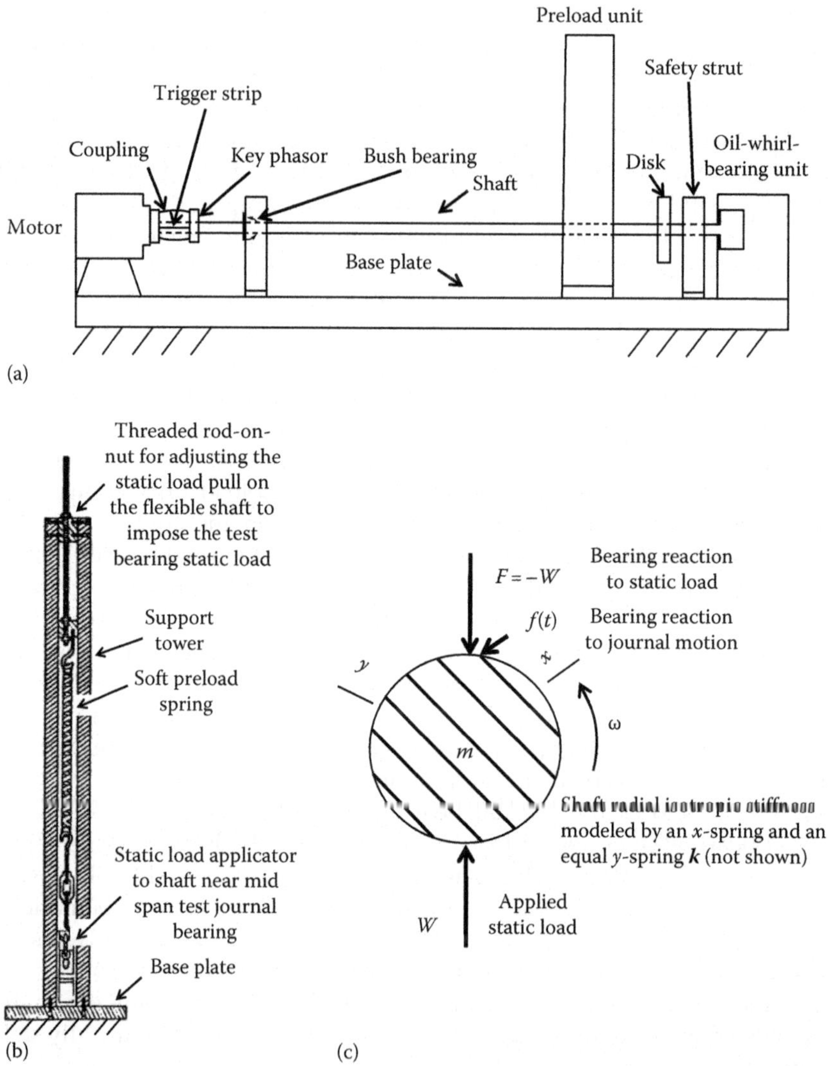

FIGURE 2.7
B-N rotor kit modified by adding a midspan hydrodynamic bearing. (a) Rig configuration, (b) bearing preload unit, and (c) simulation model.

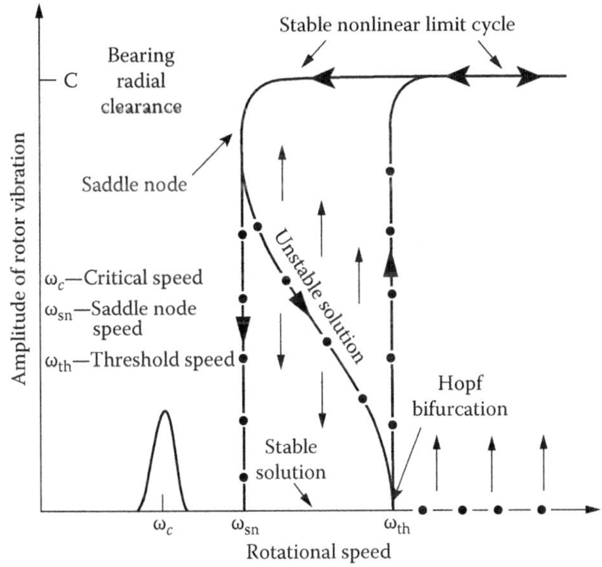

FIGURE 2.8
Journal bearing hysteresis loop.

2.4 Rotor with Two Planes of Automatically Controllable Balancers

The test rig shown in Figure 2.9 was developed to be the first implementation of a new real-time automated 2-plane rotor balancing system by Lord Corporation. It employed two rotor-mounted balancing disks controlled in real time to adjust balance correction weights to minimize residual rotor vibration as per a user-specified maximum allowable rotor vibration. Details of this research are given by Falah (2002) and Adams and Falah (2004). They also show their comparisons between the test results and theoretical analysis predictions (using the Rotor Dynamic Analysis code RDA supplied by Adams (2010)), for critical speed and oil whip threshold speed predictions.

A detailed cutaway of the Lord rotor mass balancer is shown in Figure 2.10. The rotor-mounted portion houses two equally unbalanced counterweight/stepping-motor rotors separately indexed in 5° increments relative to the rotor. Power and control are through magnetic couplers. In addition to the primary purpose of real-time automatic rotor balancing, the control system also allows the manual inputting of controlled unbalances. That option was used to adequately excite rotor unbalance-excited resonances to accurately detect critical speeds.

Conventional rotor-mounted automatic balancing devices are designed to minimize residual rotor mass unbalance so that the rotor vibration level

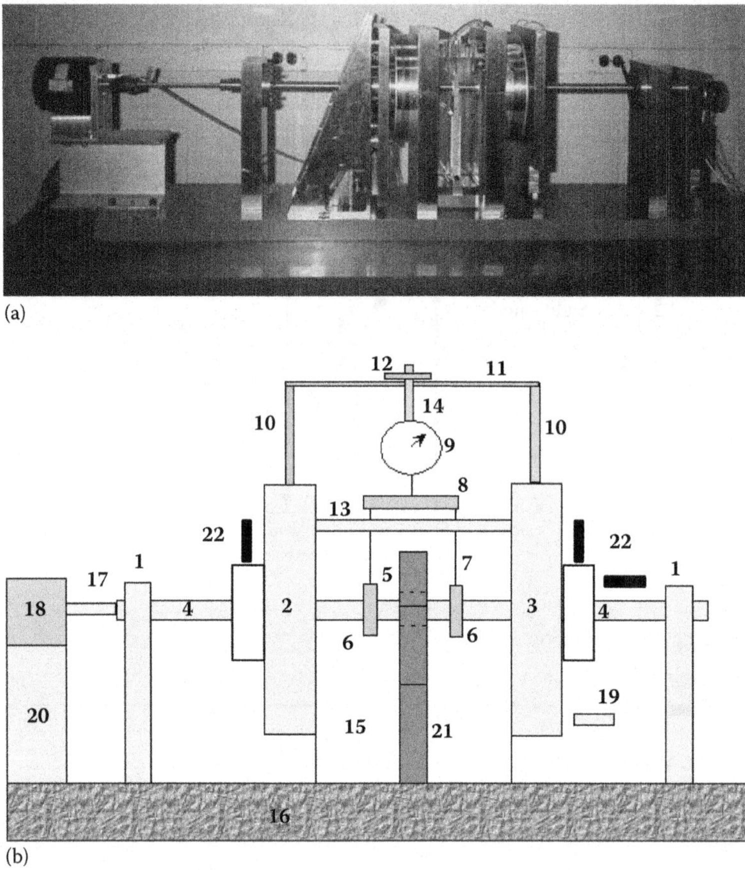

FIGURE 2.9
Rotor rig with midspan journal bearing and two Lord balancers. *Notes:* (1) End-bearing (preloaded duplex ball), (2) drive-end balancer, (3) outboard balancer, (4) shaft, (5) hydrodynamic journal bearing, (6) static load applicator, (7) lift load link, (8) lift load beam, (9) static load measurement, (10) support columns, (11) load support beam, (12) threaded load-control knob, (13) transparent cover lid, (14) threaded load rod, (15) transparent oil tank, (16) table, (17) quill shaft, (18) DC motor, (19) key phasor, (20) motor support base, (21) journal bearing support, and (22) proximity probes.

is maintained within a given application's requirements. Precision machine tool spindles, especially for grinding, are a major application for such devices since successful high-volume high-precision grinding requires continual automatic adjustment of balance correction weights on the rotating assembly as grinding wheel material is removed by wear.

Conventional devices accomplish balancing through many successive changes to the balance weight and phase angle but without "knowing" the magnitude and angular location of the continuously changing correction weight and angular location. The Lord balancer shown in Figure 2.10 has

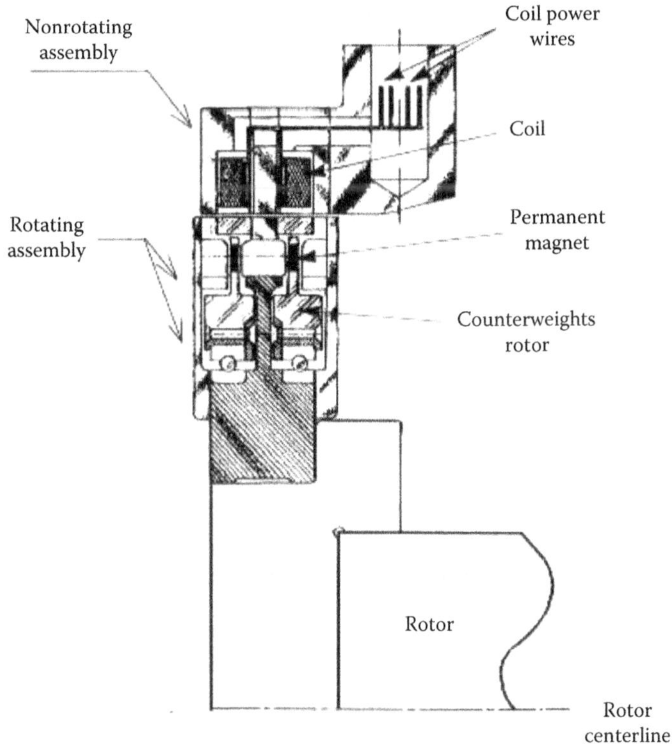

FIGURE 2.10
Lord Corporation automatic rotor mass balancer.

significantly advanced the field of automatic rotor balancing by tracking the magnitude and angular location of the instantaneous correct weights. That real-time information enables better trending in the condition monitoring of running machines. Furthermore, being able to impose a known incremental balance weight change and then tracking the accompanying change in rotor vibration signature provides invaluable diagnostic information, that is, by *dynamic probing* (Adams and Loparo 2000).

Bibliography

Adams, M. L., Non-linear dynamics of flexible multi-bearing rotors, *Journal of Sound and Vibration*, 71 (1), 129–144, 1980.

Adams, M. L., Development of advanced rotor-bearing systems for feedwater pumps—Phase III: Hardware design and fabrication, EPRI Final Report CS-3203, EPRI Project 1884-4, Electric Power Research Institute, Palo Alto, CA, 45pp., July 1983.

Adams, M. L., Insights into linearized rotor dynamics, Part-2, *Journal of Sound and Vibration*, 112 (1), 97–110, 1987.

Adams, M. L., Experiments and calculations on fundamental nonlinear rotordynamical systems, MS thesis, Case Western Reserve University, Oxford University, Oxford, U.K., p. 152, 1996.

Adams, M. L., *Rotating Machinery Vibration: From Analysis to Troubleshooting*, 2nd edn., Taylor & Francis, CRC Press, Boca Raton, FL, 465pp., 2010.

Adams, M. L., *Power Plant Centrifugal Pumps: Problem Analysis and Trouble-Shooting*, Taylor & Francis, CRC Press, Boca Raton, FL, 182pp., 2017.

Adams, M. L., Adams, M. L., and Guo, J. S., Simulations & experiments of the nonlinear hysteresis loop for rotor-bearing instability, *Proceedings of Sixth IMechE International Conference on Vibration in Rotating Machinery*, Oxford University, Oxford, U.K., Sept. 1996.

Adams, M. L., Afshari, F., and Adams, M. L., An experiment to measure the restitution coefficient for rotor-stator impacts, *Proceeedings of Seventh IMechE International Conference on Vibration in Rotating Machinery*, Nottingham, England, pp. 301–308, Sept. 2000.

Adams, M. L. and Falah, A. H., Experiments and modelling of a three-bearing flexible rotor for unbalance response and instability thresholds, *Proceedings of Eighth IMechE International Conference on Vibration in Rotating Machinery*, Swansea, Wales, U.K., pp. 595–602, 2004.

Adams, M. L. and Loparo, K. A., Model-based condition monitoring from rotating machinery vibration, Final Report, EPRI Project WO3693-04, Electric Power Research Institute, Palo Alto, CA, 2000.

Adams, M. L. and McCloskey, T. H., Large unbalance vibration in steam turbine-generator sets, *Vibrations in Rotating Machinery*, Institution of Mechanical Engineers, York, England, pp. 491–497, 1984.

Adams, M. L., Sawicki, J. T., and Capaldi, R. J., Experimental determination of hydrostatic journal bearing rotordynamic coefficients, *Proceedings, Fifth IMechE International Conference on Vibration in Rotating Machinery*, Bath, England, pp. 365–374, Sept. 1992.

Adams, M. L., Yang, T., and Pace, S. E., A sea rest facility for the measurement of isotropic and anisotropic linear rotordynamic characteristics, *Proceedings of NASA Sponsored Workshop on Rotordynamic Instability Problems in High-Performance Turbomachinery*, Texas A & M University, College Station, TX, NASA CP-3026, May 1988.

Chamieh, D., Acosta, A. J., and Caughey, T. K., Experimental measurements of hydrodynamic stiffness matrices for a centrifugal pump impeller, *Workshop: Rotordynamic Instability Problems In High Performance Turbomachinery*, Texas A & M University, College Station, TX, NASA CP No. 2250, May 1984.

Chamieh, D., Caughey, T. K., Brennen, C. E., and Acosta, A. J., Comments on impeller-volute interactions, *Proceedings, Power Plant Feed Pumps-State of the Art, EPRI Symposium*, Cherry Hill, NJ, June 1982.

Childs, D., Private communications with M. L. Adams, *Fifth IMechE International Conference on Vibration in Rotating Machinery*, Bath, England, Sept. 1992.

Falah, A. H., Modeling and experiments of linear and nonlinear dynamics of a flexible multi-bearing rotor, PhD thesis, Case Western Reserve University, May 2002.

Goldsmith, G., *Impact: The Theory and Physical Behavior of Colliding Solids*, Edward Arnold Publishers, Ltd., London, U.K., 1960.

Guelich, J. F., Bolleter, U., and Simon, A., Feedpump operation and design guidelines, EPRI Final Summary Report TR-102102, Research Project 1884-10, Electric Power Research Institute, Palo Alto, CA, 1993.

Horattas, G. A., Experimental investigation of dynamic nonlinearities in rotating machinery, PhD thesis, Case Western Reserve University, Aug. 1996.

Horattas, G. A., Adams, M. L., and Dimoftte, F., Mechanical and electrical run-out removal on a precision rotor-vibration research spindle, *ASME Journal of Acoustics and Vibration*, 119 (2), 216–220, 1997.

Hori, Y. and Kato, T., Earthquake induced instability of a rotor supported by oil film bearings, *ASME Journal of Vibrations and Acoustics*, 112, 160–165, 1990.

Makay, E., Survey of feed pump outages, EPRI Research Project RP 641, Final Report FP-754 (Edited by M. L. Adams), Electric Power Research Institute, Palo Alto, CA, 98pp., 1978.

Makay, E., How close are your feed pumps to instability-cased disaster, *Power Magazine*, 69–71, 1980.

Makay, E. and Barrett, J., Changes in hydraulic component geometries greatly increased power plant availability and reduced maintenance costs: Case histories, *Proceedings, Texas A & M First International Pump Symposium*, Texas A & M University, College Station, TX, 1984.

Quinn, R. D., Experimental study of uncentralized squeeze film dampers, MS thesis, University of Akron, p. 117, 1984.

Sawicki, J., Capaldi, R. J., and Adams, M. L., Experimental and theoretical rotordynamic characteristics of a hybrid journal bearing, *ASME Journal of Tribology*, 119 (1), 132–142, Jan. 1997.

3

Large Steam Turbine Generator Turning-Gear Slow-Roll Journal Bearing Load Capacity

A major journal bearing investigation conducted by the author in the 1970s was focused upon experimentally determining journal bearing load capacity for large U.S.-manufactured main steam turbine generator sets immediately following shutdown when turbine generator units are put on turning-gear-driven slow-roll, 3–5 rpm. Because these turbine units are quite hot when first shut down, they are not immediately brought down to 0 rpm to avoid thermal bowing of the rotors as they cool down.

3.1 Prior State-of-the-Art and Motivations

To operate successfully under turning-gear speed, the long-standing design guideline was to size these journal bearings for a nominal *Unit Load** of P = 200 psi (13.6 bar), *$P = W/DL$, where W is the bearing static load and diameter D times length L is the projected area of the bearing cylindrical surface. But there was no documented evidence as to the origin or basis for this bearing design load limitation.

As a consequence, the journal bearings were typically oversized for running at the operating speed of 3600 rpm, 2-pole synchronous speed for a 60 Hz generator. As a consequence of somewhat oversized bearings, many of the low-pressure (LP) steam turbines were operating just below the oil whip threshold speed. This was evidenced by an occasional oil whip vibration occurring on an operating unit in the field, requiring a field fix for that unit to eliminate the large-amplitude subsynchronous rotor vibration characteristic of oil whip (see Section 2.3). When oil whip vibration occurs, its vibration level typically far exceeds the maximum allowable vibration level. Thus, the unit cannot be operated until the oil whip is eliminated through field adjustment, such as realigning the multibearing elevations to add static load to the offending bearing.

Clearly, the major motivation for this research was the possibility of eventually increasing the design unit load on large turbine generator journal bearings. The shaft journal diameters of these high-power machines must be sized to carry the high transmitted torques through the rigidly coupled multi-turbine driveline with the last LP turbine's end journal transmitting the full power torque input to the generator. Therefore, increasing the design unit load on the journal bearings can only be done by reducing the bearing's active length *L*.

The advantages to be gained by increasing journal bearing design unit loads, that is, designing the bearings smaller, include (1) greater assurance of stable oil whip-free rotor-bearing dynamics, (2) reductions in bearing power losses at normal running speed with the accompanying benefit of lower-capacity auxiliary equipment, and (3) the reduction of journal diameters in applications where the journals are not already torque limited.

3.2 Test Rig Design

The broad scope of tests deemed necessary dictated that the size of bearings tested be limited in diameter—3 inches (7.6 cm) being accordingly chosen. Tests were conducted with five (5) different speeds, four (4) different oil temperatures, three (3) different babbitt materials, three (3) different journal materials, two (2) different babbitt thicknesses, and various bearing loads. Each test (61 in total) required a new test bearing because wear phenomena were involved. Such a large scope of testing with, say, a 16-inch (41 cm journal diameter (typical LP turbine journal) would not have been economically feasible. The test rig designed by the author for this investigation enabled the broad test scope, being configured for rapid change of bearing and journal specimens and operation unattended overnight.

Figure 3.1 shows a schematic of the entire rig configuration. Journal speeds of 1.5–50 rpm were achieved with a gear reduction unit driven by a positive-displacement hydraulic motor, which provided constant speed virtually independent of torque variations, unlike an electric motor. As the

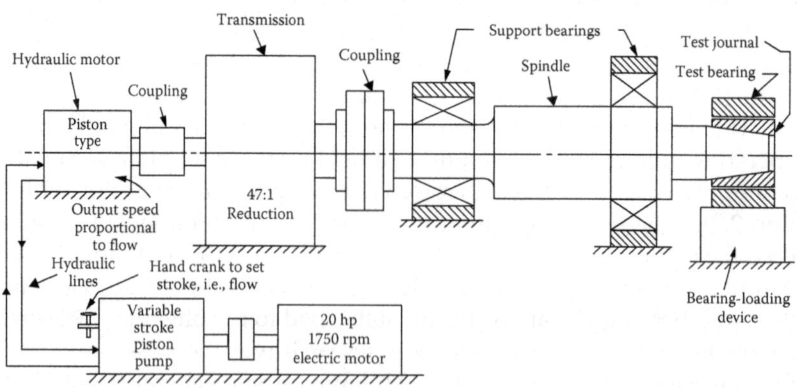

FIGURE 3.1
Slow-roll turning-gear journal bearing test rig.

test bearing drag torque changes during a test, the pressure in the positive-displacement hydraulic drive automatically adjusts to accommodate the instantaneous bearing torque. The test journal was mounted on the spindle end, employing a machine tool taper fit to facilitate journal specimen replacement without incurring significant radial runout. The overall spindle and support construction was built quite robustly to achieve very high overall stiffness to render structural deflections insignificant. The spindle and non-rotating portion of the rig were electrically insulated from each other in order to employ the *electrical resistance* method to detect when a test bearing and journal become completely separated by a lubricant film.

The axial view from the test end of the spindle illustrated in Figure 3.2 shows the test bearing fixture vertically loaded through a floating hydrostatic

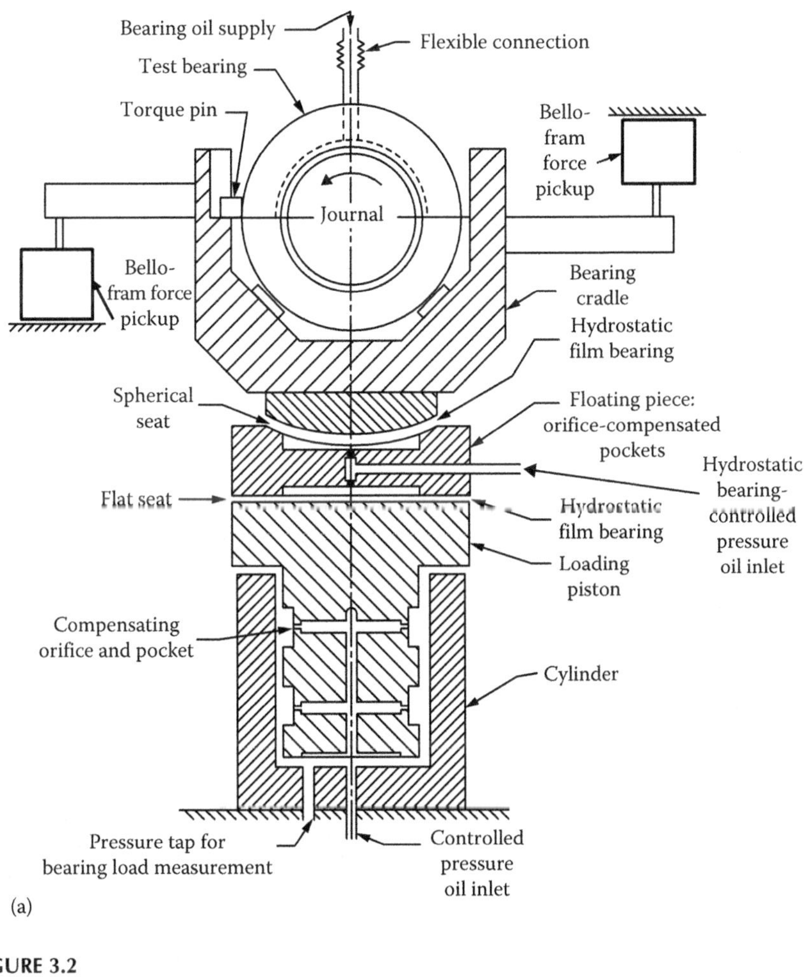

FIGURE 3.2
(a) Test bearing vertical load train. (*Continued*)

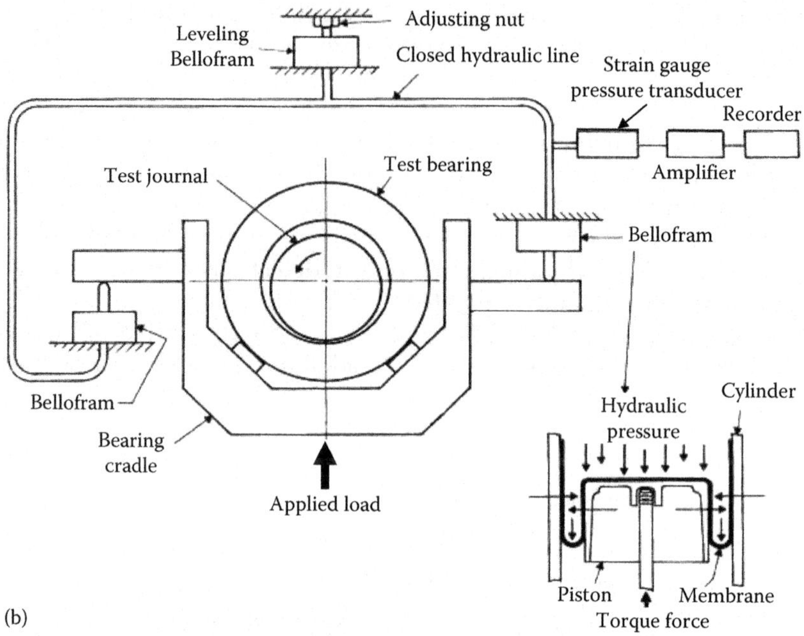

FIGURE 3.2 (*Continued*)
(b) Bearing torque measurement.

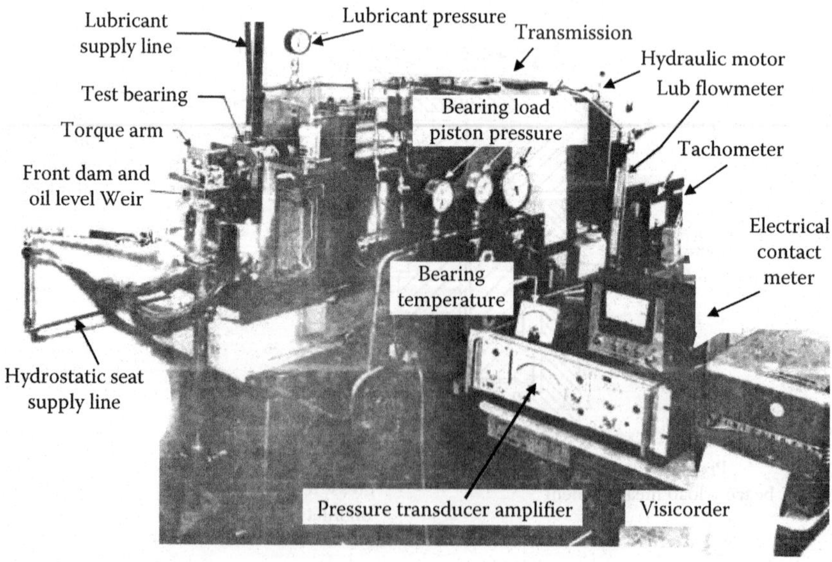

FIGURE 3.3
Turning-gear test rig and instrumentation.

thrust bearing, spherical on the top surface and flat on the bottom surface. That combination eliminates any misaligning moment or horizontal side load applied to the test bearing. The test bearing is thus loaded only by an upward vertical force, simulating a constant vertical gravitational force.

As shown in Figure 3.2(b), the bearing torque-measuring system is quite unique, keeping the bearing from angulating as test bearing torque varies considerably over the duration of a test. The two identical Bellofram sealed chambers are hydraulically dead-ended against each other, guaranteeing equal but opposite torque-measuring forces. Thus, this torque measurement does not add any extraneous radial load to the test bearing. At the test setup, the third Bellofram shown in the closed hydraulic line is plunge-position adjusted to achieve a near-perfect annular positioning of the test bearing saddle. The static pressure within the three Belloframs' closed hydraulic line calibrates linearly to the bearing-restraining torque. A photo of the complete test rig and instrumentation is shown in Figure 3.3.

3.3 Test Bearing and Journal Details

Both 180° partial-arc cylindrical bearings and the two loaded pads of 4-shoe tilting-pad journal bearings were tested, since both configurations are used on some commercially supplied main steam turbine generator sets. Figure 3.4 shows photos of both these test bearing configurations as mounted in the test rig bearing cradle. Both the bearing and journal surface finishes were in close conformance with those of the actual turbine generator units. The journal and bearing babbitt lining materials were of the precise same specifications as in the actual turbine generator units.

Early tests were run at 3 rpm, a typical turning-gear rpm on large turbine generators at journal bearing unit loads within the product's 200 psi guideline. Figure 3.5(a) and (b) show two of those bearings after testing. It was clear from the distressed appearance of those two bearings tested at 3 rpm alone that *journal rpm is not the relevant speed factor* to equilibrate slow-roll unit load capacity between journal bearings of substantially different diameters. On the other hand, the photo of the tested bearing in Figure 3.5(c) shows an excellent appearance after running at a 400 psi unit load but at 10.5 rpm, which is a journal surface speed typical of the large steam turbine generator units at 3 rpm. To further solidify that surface speed was the relevant speed factor, a second journal spindle (see Figure 3.1) was made to accommodate a nominal 1.5-inch-diameter journal and bearing. Tests with 1.5-inch-diameter bearings further confirmed that journal surface speed was the relevant speed parameter.

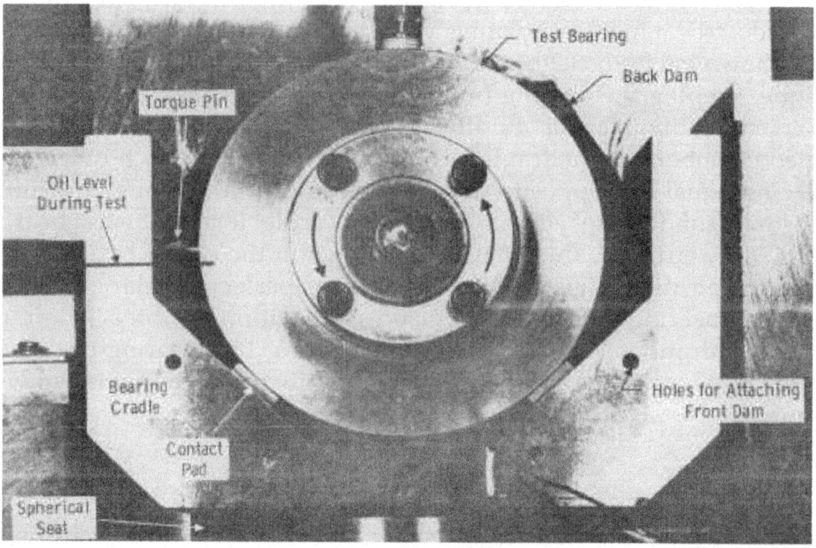

(a)

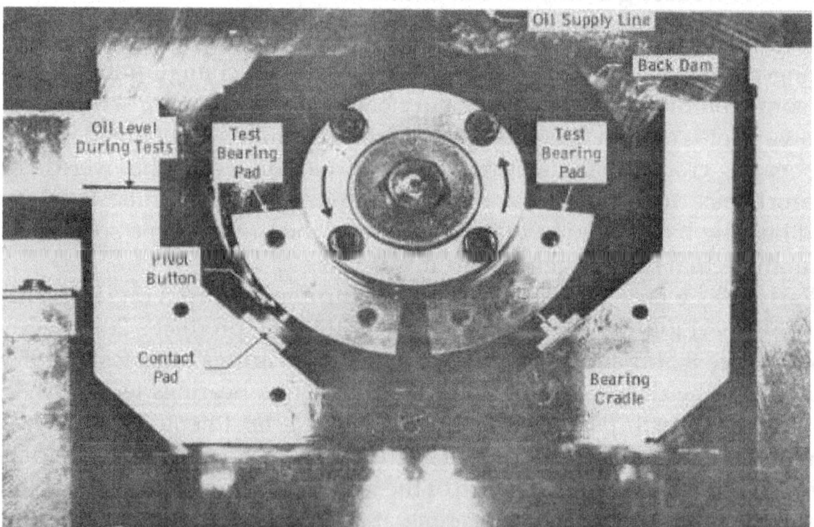

(b)

FIGURE 3.4
Test bearing configurations: (a) cylindrical sleeve and (b) tilting pad.

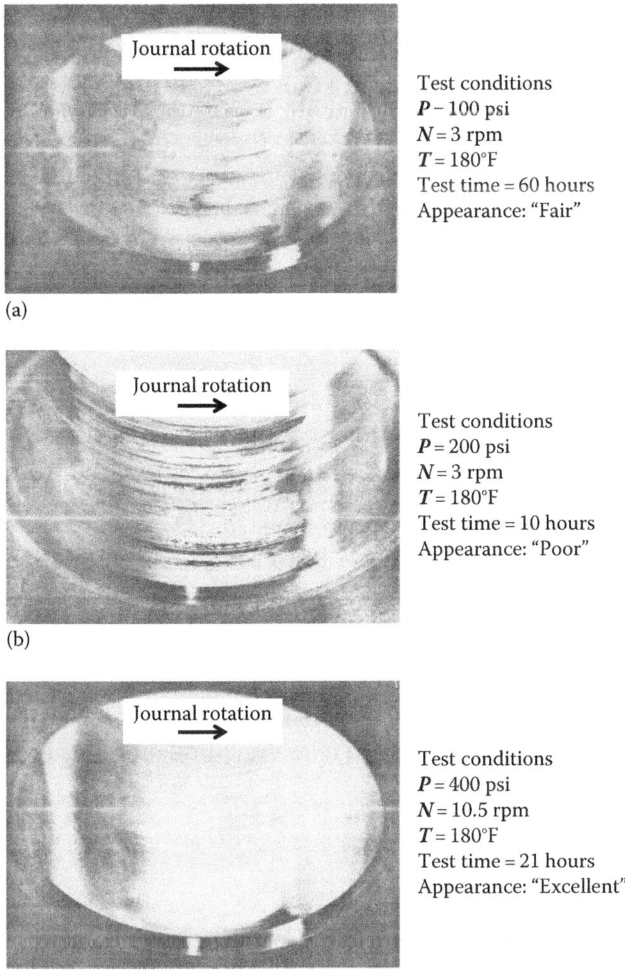

FIGURE 3.5
Photos of tested bearings indicating that surface speed is the relevant speed: bearing specimens (a), (b), and (c) all of the identical 3 inch diameter and 2.35 inch length and babbitt thickness.

3.4 Test Duration and Measurements

When starting a large main steam turbine generator unit from a cold zero-speed shutdown, the journal bearings, of course, start under full static load, that is, the full weight of the rotor driveline. So tests were started (time $t = 0$) by turning on the spindle drive at the designated test speed with the bearing already under the designated test load. The bearing torque was continuously measured from time = 0 and throughout the complete duration of the test—nominally 24 hours. The torque force measurement signal was recorded with a strip recorder (Visicorder) employing a fast-acting optical beam. The recorder was run at its highest strip speed (50 inches per second) for the first few seconds of the test to capture the breakaway friction transient. Keep in mind that high-speed real-time A-to-D data acquisition systems were not available in the 1970s. Figure 3.6 is a typical example of the bearing breakaway torque transient scaled to the coefficient of friction.

The history of the friction coefficient as a test progresses in time is one of the major measurements used in judging whether a given test condition exhibits an *acceptable* or *unacceptable* operating condition. Figure 3.7 shows and tabulates the criteria upon judging two contrasting test results. As tabulated, important contributing symptoms in assessing the *acceptable* or *unacceptable* nature of the operating condition include (1) the friction coefficient test record, (2) electrical resistance measurements, and (3) the after-test appearance of the test bearing.

As mentioned in Section 3.2, the spindle and nonrotating portion of the rig were electrically insulated from each other in order to employ the *electrical resistance* method to detect when a test bearing and journal become completely separated by a lubricant oil film. Very thin lubricating films were not directly measurable thicknesswise. Consequently, the electrical resistance method had long been used by friction-and-wear researchers to experimentally detect when a marginally lubricated interaction between the asperities of two contacting surfaces in relative sliding motion involves metal-to-metal contact between the two surfaces in relative sliding motion. Figure 3.8 shows

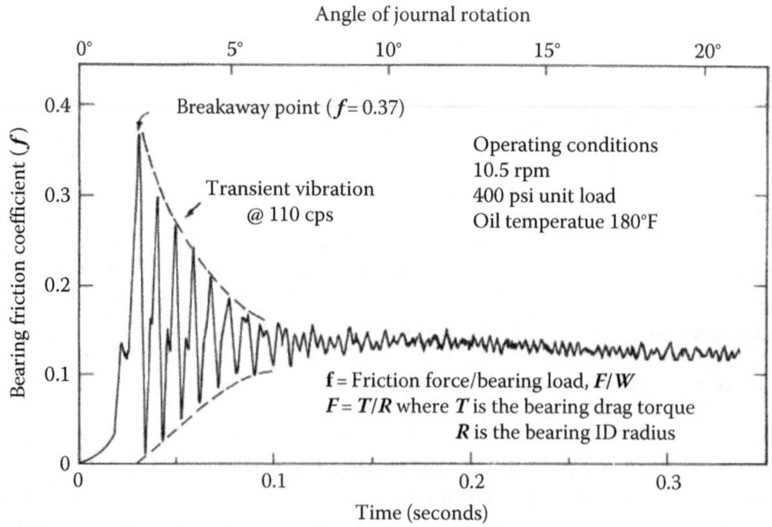

FIGURE 3.6
Test startup Visicorder trace scaled to bearing coefficient of friction f.

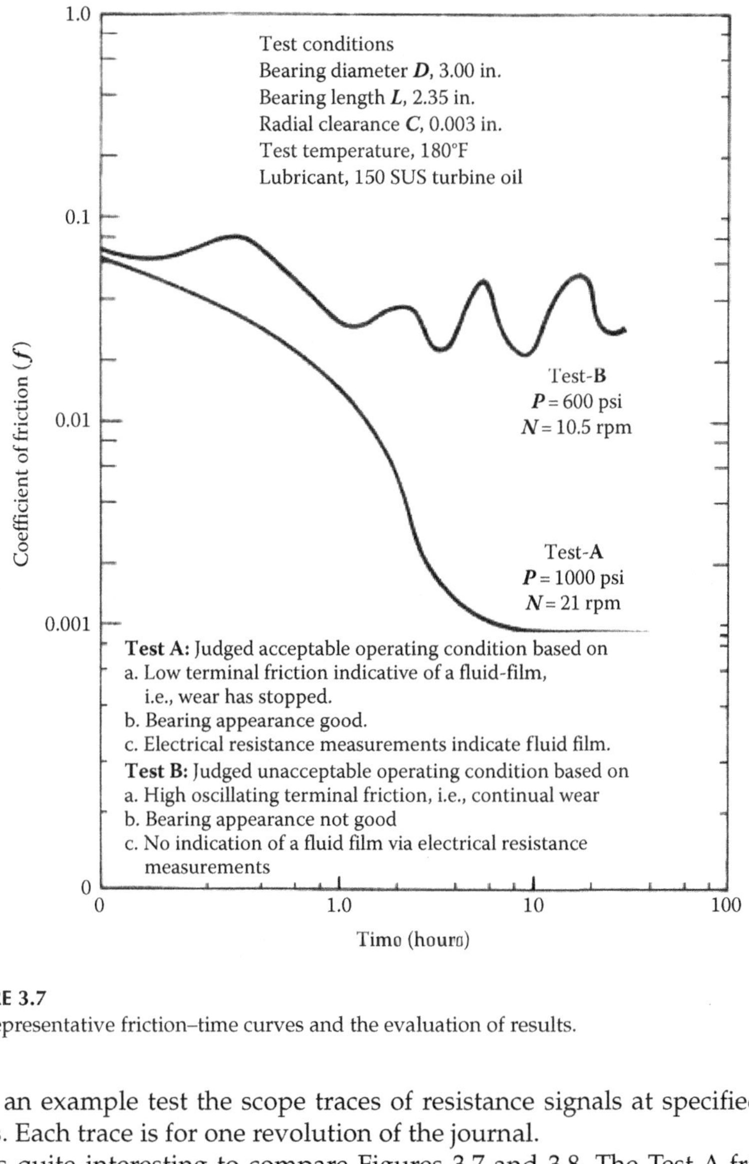

FIGURE 3.7
Two representative friction–time curves and the evaluation of results.

from an example test the scope traces of resistance signals at specified test times. Each trace is for one revolution of the journal.

It is quite interesting to compare Figures 3.7 and 3.8. The Test-A friction versus time curve indicates a *successful wear-in* after about 10 hours of running, whereas the electrical resistance scope traces in Figure 3.8 show that the complete separation of a journal and bearing by a lubricant film did not occur until a much longer test time (75.8 hours). Clearly, the electrical resistance signal is an even more stringent indicator of whether or not two rubbing surfaces are separated by a lubricant film.

A high-precision Gould surface-trace analyzer was employed to record bearing and journal surface finishes and waviness both before and after a

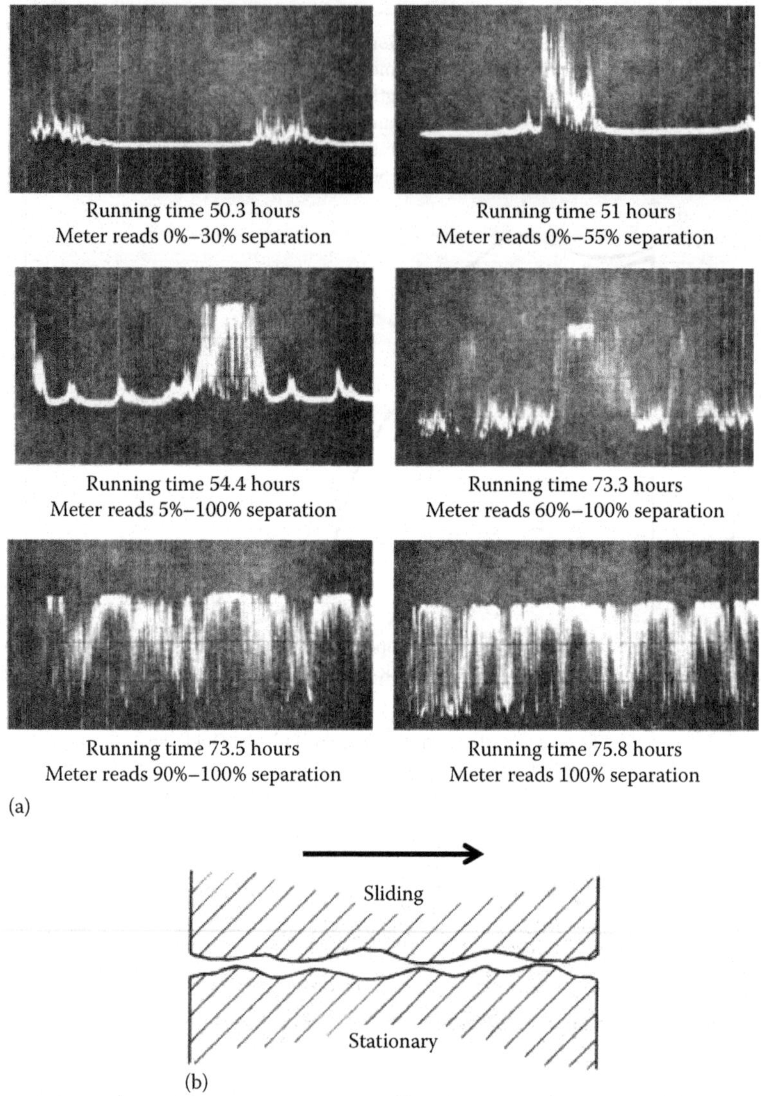

FIGURE 3.8
(a) Scope traces of electrical resistance at bearing–journal contact and (b) real contact surfaces of smooth solids (asperities exaggerated).

test. The Gould surface analyzer was equipped with both a straight-line tracer and a rotating-platform tracer. So the pretest and posttest surface traces were used to measure bearing circular and axial wear patterns. Some of the tests exhibiting a Test-A (Figure 3.7)-type acceptable wear-in leading to lubricate film separation of the bearing and journal were installed back into the test rig to see if a second but much less wear-in would occur. Wear measurements with the Gould surface analyzer confirmed that there was no significant second wear-in. It is worth mentioning here that since the entire turbine generator is axially fixed by a thrust bearing, the differential axial thermal expansion between the rotor and bearings are significant. Thus in the field, there surely occurs small secondary wear-ins. Fortunately, these large generating machines are not regularly brought offline and shut down since they are designed to run 24/7 to continually produce electric power.

3.5 What Tests Revealed

With test specimen surface finish and waviness, materials, operating temperatures, journal surface speeds (not rpm), and lubricant of the actual turbine generators, bearing unit loads far in excess of the 200 psi design guideline were easily supported at turning-gear journal surface speeds. The test results therefore suggest that unit load capacity could be safely increased above the 200 psi level. A test with bearing unit load up to 1600 psi was carried with no resulting distress. And in that test, seven contact-arc pressure-taps confirmed a corresponding generated hydrodynamic film pressure, Figure 3.9. Turbine generator journals' active diameter D must be sized to handle the torque of transmitted power, so increasing bearing unit load entails reducing the bearing active length L.

When a new journal bearing is first operated on turning-gear under load, a slight wear-in of the babbitt surface takes place typical of an abrasive wear phenomenon where the harder shaft ploughs the much softer babbitt. Bearing surface finish is thereby much improved. After wear-in, the friction is very low, indicative of the presence of a very thin hydrodynamic oil film, rendering subsequent babbitt wear negligible. The breakaway friction coefficient is not a direct function of applied load, consistent with the long-standing simple adhesive-wear-based Coulomb friction approximation. The range of breakaway **f** was found to be 0.22–0.43.

Turning-gear speed bearing load capacity is roughly proportional to journal surface speed, lubricant viscosity, and clearance. Thus, 3 rpm permits twice the unit load capacity as 1.5 rpm. Low-speed bearing load capacity increases considerably with reduced journal roughness.

Super finishing of journals would increase low-speed bearing reliability. Journals in the field sometimes acquire surface scratches due to the ingestion of hard dirt particles within the oil. Testing with imposed circumferential journal grooves did not noticeably change low-speed load capacity. Not surprisingly, however, testing with an imposed axial groove seriously damaged the bearing babbitt surface.

Babbitt liner thickness is not a significant parameter affecting low-speed bearing load capacity. However, a tin-based babbitt is considerably better

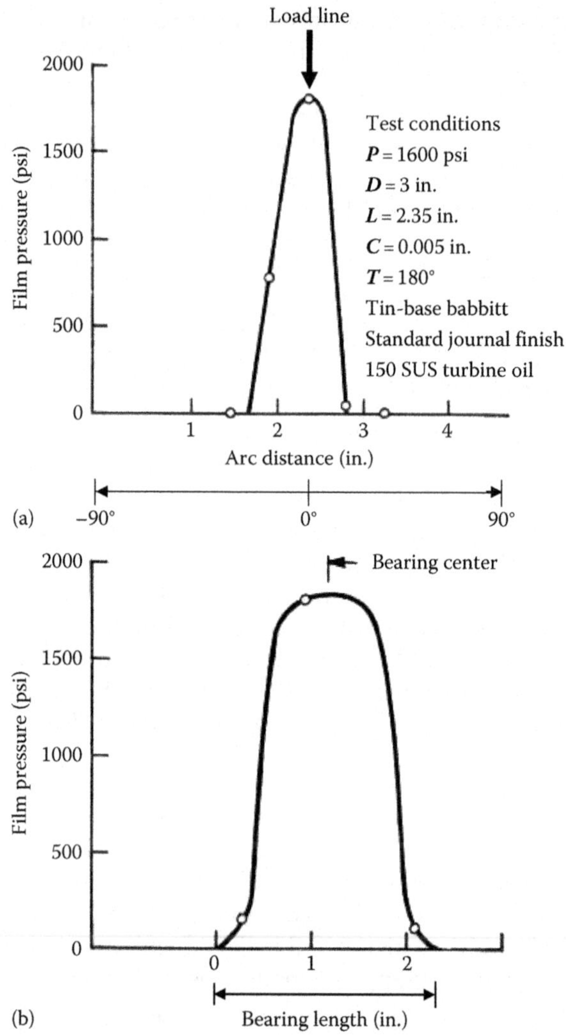

FIGURE 3.9
Lubricant film pressure measurements: (a) circumferential and (b) axial.

than a lead-based babbitt. All of the tests except two were run with tin-based babbitt. The bearing L/D ratio was not found to affect low-speed unit load (P) capacity.

The low-speed capacity of a tilting-pad bearing is not significantly different than an equivalent fixed partial-arc configuration, but perhaps a little better, given that the bearing load is shared by two pads (arcs).

4

Journal Bearing and Radial Seal Rotor Dynamics

Chapter 1 covers the double-spool spindle test rigs of both the Case Western Reserve University (CWRU) machinery dynamics laboratory and the Caltech centrifugal pump laboratory. The CWRU rig was developed for journal bearing and radial seal rotor dynamics properties, and for rub-impact characteristics, all for EPRI- and NASA-sponsored research projects. The Caltech rig was developed for NASA-sponsored research to study hydraulic static and dynamic forces in high-speed centrifugal pumps, including the rotor dynamic property measurement for a complete centrifugal pump stage. This chapter covers a number of other different test rig types that were developed specifically to measure the rotor dynamic properties of journal bearings and radial seals. The author (2010) presents an extensive background on the importance of rotor–stator dynamic interactions from bearings, seals, and turbomachinery stages. The emphasis in this chapter is on test rig types that have been developed specifically to research bearing and radial-seal rotor dynamic properties. Additional test rigs used in this research area are also presented in Chapter 18, which is committed specifically to the TAMU turbomachinery laboratory.

4.1 Mechanical Impedance Method with Harmonic Excitation

Impedance approaches are often associated with the characterization of an electrical network by a prescribed model circuit of resistances, capacitances, and inductances. With sufficient measured input/output data on an actual system, a correlation of the *input* (e.g., single-frequency sinusoidal voltage signal) and the resulting *output* (e.g., current signal) leads to a solution of values for all the model's resistances, capacitances, and inductances that would theoretically produce the measured outputs caused by the measured inputs. Such a characterization process is commonly referred to as *system identification*. Quite similar approaches have long been used to characterize mechanical dynamic systems with a suitable linear model in which the values of a discrete model's stiffness, damping, and mass elements are solved by determining what combination of these values produces the "best fit" in correlating measured input and output signals, force and motion or vice versa.

For example, suppose a machine is mounted on the floor of a large plant and it is known from experience that if a vibration analysis of the machine assumes the floor to be perfectly rigid, the analysis will be seriously flawed. Common sense dictates that one does not devise a finite-element model of the entire plant building just to couple it to the vibration model of the machine in question. If previous experimental data is not deemed applicable, then a mechanical impedance *harmonic shaker test* can be performed on the plant floor at the location where the machine will be installed.

An alternate technique is to apply an impact force to the floor position in question, measuring simultaneously the impact force signal and the acceleration signal at the floor point of impact (specific example, test rig described in Section 4.2, next). Impact approaches are fairly common and standard *hammer kits*, from small laboratory size up to large sledgehammer size for power plant machinery, are made for this purpose with the force and motion signals processed through a dual-channel FFT instrument to extract the impact point's mechanical impedance. For very large structures (e.g., a plant building) or devices with very high internal damping (e.g., a multistage centrifugal pump), single-impact techniques may lack sufficient energy input to the structure to achieve adequately high response signal-to-noise ratios to work well. In such cases, multiple impact strikes, for example, several hundreds, combined with synchronized signal time-averaging have been used to filter out the noncoherent signal noise, but this is a very specialized procedure. For the sake of the following example, it is assumed that the vertical motion of the floor is significant and that a mechanical shaker is used as illustrated in Figure 4.1.

If the immediate structure is dynamically close to linear, then its steady-state response will significantly contain only the forcing-function frequency component ω. The linearity assumption leads to the following equations as

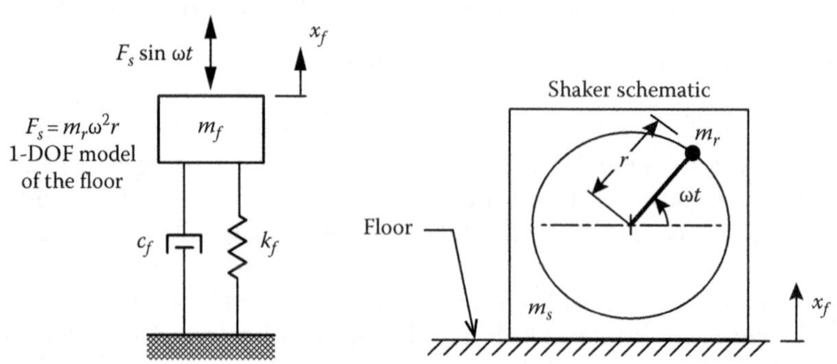

FIGURE 4.1
Vertical shaker test of a floor where a machine is to be installed.

the basis for processing measured response to the controlled sinusoidal force input illustrated in Figure 4.1. It is convenient to use the complex plane representation for the harmonic input and output response signals:

$$\begin{aligned}(m_s + m_f)\ddot{x}_f + c_f\dot{x}_f + k_f x_f &= F_s e^{i\omega t}\\ x_f &= Xe^{i(\omega t+\phi)}\end{aligned} \tag{4.1}$$

Equations 4.1 lead to the following complex algebraic equation:

$$\left(k_f - \omega^2 m_f - \omega^2 m_s + ic_f\omega\right)Xe^{i\phi} = F_s \tag{4.2}$$

For the configuration illustrated in Figure 4.1, the vertical forcing function is equal to the imaginary part of the complex force. The single complex algebraic Equation 4.2 is equivalent to two real algebraic equations and thus can yield a solution for the two unknowns, $(k_f - \omega^2 m_f)$ and c_f, at a given frequency, where shaker mass m_s is known *a priori*. If the excited floor point were in fact an exact 1-Degree-of-Freedom (1-DOF) system, its response would be exactly that of the 1-DOF system and the impedance coefficients k_f, m_f, and c_f would be constants independent of vibration frequency, ω. However, since an actual building structure is sure to be dynamically far more complex than a 1-DOF system, the "exact-fit" impedance coefficients will be functions of frequency. In that case, when it is deemed appropriate or necessary to treat the impedance coefficients as "constants" over some frequency range of intended application, the coefficients are then typically solved using, for example, a *least squares linear regression* fit of measurement data over the applicable frequency range.

As presented in Section 1.1.1, for a 2-DOF radial plane motion experiment on a dynamically anisotropic bearing or radial seal, force and motion signals must be processed in two different radial directions, preferably orthogonal like the standard x-y coordinates. For a near-concentric fluid annulus, such as that typically assumed for radial seals, the dynamic coefficient matrices are formulated to be isotropic, as is consistent with a rotationally symmetric steady-state flow field. Impedance tests devised strictly for the isotropic model require fewer data signals than impedance tests for the anisotropic model (Adams 2010).

There are fundamentally two ways of designating the harmonic *inputs* and *outputs*. In the 1-DOF impedance test schematically illustrated in Figure 4.1, the input is the harmonic force and the output is the resulting harmonic displacement response. However, there is no fundamental reason that prevents these roles from being reversed, since both input and output signals are measured and then likewise processed through Equation 4.1. Likewise, a 2-DOF radial plane motion experiment on a bearing or radial seal may have the x and y force signals as the controlled inputs with the resulting x and y displacement signals as the outputs or the converse of this. Both types of test methods are used for bearing and radial seal dynamic characterizations. The test rig covered in Section 1.1 employs the prescribed circular harmonic orbital motion (x, y) as the 2-DOF inputs and the resulting 2-DOF harmonic force vector (F_x, F_y) as the outputs. In that arrangement, the four Equation 1.3 applied at three different frequencies provide the 12 linear algebraic equations from which to solve for the 12 unknowns, that is, the three 2×2 stiffness, damping, and inertia matrices shown in Equation 1.1.

4.2 Mechanical Impedance Method with Impact Excitation

As explained earlier in Section 4.1, impact techniques are widely used on lightly damped structures with relatively low background structural vibration noise levels. Such structures can be impact-excited to yield adequately favorable signal-to-noise ratios. Journal bearings and fluid-annulus seals typically have considerable inherent damping, and that is always an important benefit for controlling rotor vibration to within acceptable residual vibration levels in operating machinery. However, from the point-of-view of using impact testing to extract dynamic coefficients of bearings and other fluid-annulus elements, their inherent damping capacity typically results in unfavorable signal-to-noise ratios. Nevertheless, the relative ease of conducting impact impedance testing motivated some researchers to pursue the impact impedance approach for various rotor–stator fluid-annulus elements. An example is the work of Nordmann and Massmann (1984) of Germany, whose experimental setup is illustrated in Figure 4.2.

Nordmann and Massmann demonstrated the use of single-impact testing to extract the stiffness, damping, and inertia rotor dynamic coefficients for radial seals. For coefficient extraction, they used the isotropic model and also correctly conformed to the author's (1987, 2010) axiom that physically consistent modeling requires symmetry on the highest-order coefficient array, that is, requiring $m_{xy} = m_{yx} = 0$ when fluid inertia is included. Their test setup employs two "identical" test seals that are fed from a common central pressurized annular chamber and are axially opposed to cancel axial pressure forces on the quite flexibly supported seal housing. The housing can be impacted at its center-of-mass in various x–y radial directions, so a 2-DOF x-y model is thereby used as the basis for processing the measured signals to extract the five coefficients of the physically consistent isotropic model. Various algorithms, such as least-squares linear-regression fitting, can be employed to extract the five isotropic-model dynamic coefficients to provide

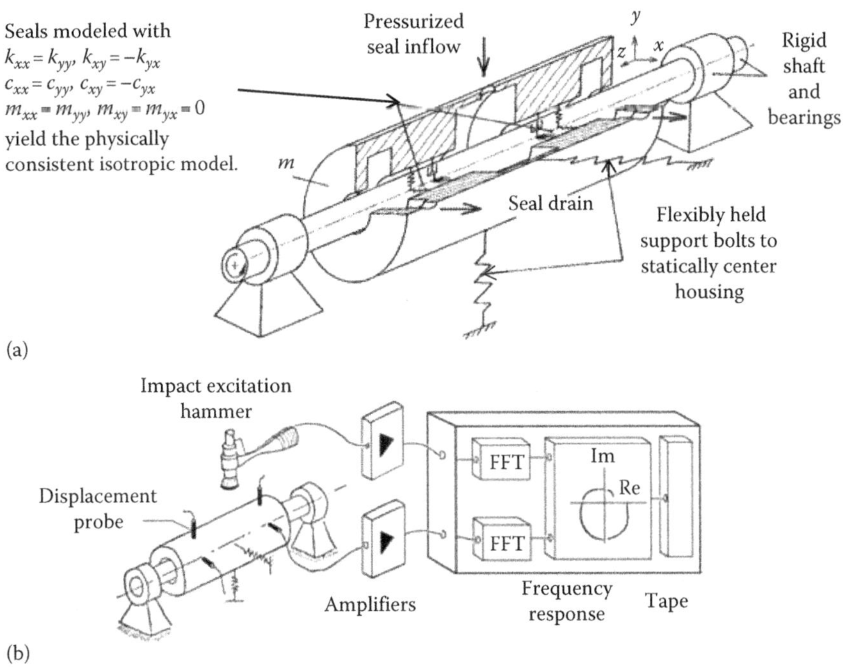

FIGURE 4.2
Test rig for a single-impact excitation of radial seals: (a) quarter through-cut schematic of test apparatus and (b) schematic of test measurements and data processing.

the "best" frequency response fit of the model to the measured time-based signals as transformed into the frequency domain. The 2-DOF model's equations of motion are as follows with the factor of "2" present because the apparatus has two "identical" radial seals.

$$
\begin{aligned}
m\ddot{x} + 2\left(m_{xx}\ddot{x} + c_{xx}\dot{x} + c_{xy}\dot{y} + k_{xx}x + k_{xy}y\right) &= F_x(t) \\
m\ddot{y} + 2\left(m_{yy}\ddot{y} + c_{yy}\dot{y} - c_{xy}\dot{x} + k_{yy}y - k_{xy}x\right) &= F_y(t)
\end{aligned}
\quad (4.3)
$$

The author suspects that a fundamental shortcoming of their apparatus and results is the inherent difficulty of getting sufficient energy from the single hammer impact into a system that has significant internal damping. Multiple impact strikes, for example, several hundreds, combined with synchronized signal time-averaging, have been successfully demonstrated by Marscher (1986) on power-plant-operating centrifugal pumps to filter out the noncoherent signal content and to obtain a pump's actual natural frequencies and corresponding mode shapes while at running conditions of head, flow, and speed.

4.3 Oil Whip Instability-Threshold-Based Rig

As explained by the author (2010), providing accurate damping inputs to vibration analysis models is possibly the most elusive aspect of making reliable predictions of vibration characteristics for almost any vibratory system. The mass and flexibility characteristics of typical structures are now accurately obtained using modern Finite Element Analysis modeling. Thus, natural frequencies and their corresponding natural mode shapes can be predicted with good accuracy in most circumstances. On the other hand, predicting the vibration *amplitude* of a *forced resonance* at a natural frequency is not such a sure thing because of the inherent uncertainties in damping inputs to the computation model. Similarly, predictions of *self-excited vibration instability thresholds* also suffer from lack of high reliability for the same reason—inherent uncertainties in damping inputs. Motivated by this fundamental reality in vibration analyses for important rotor vibration predictions, Adams and Rashidi (1985, 1988) devised a novel experimental approach for extracting bearing damping coefficients from measurements at instability thresholds. Their proposed experiment is embodied in the apparatus concept shown in Figure 4.3.

The apparatus illustrated was first proposed by Adams and Rashidi (1985). The fundamental concept behind the approach is to capitalize on the physical requirement for an exact *energy-per-cycle* balance between *positive and negative damping* influences right at an instability threshold speed. As the author (2010) details for the linear model, the complete *nonconservative portion* of the

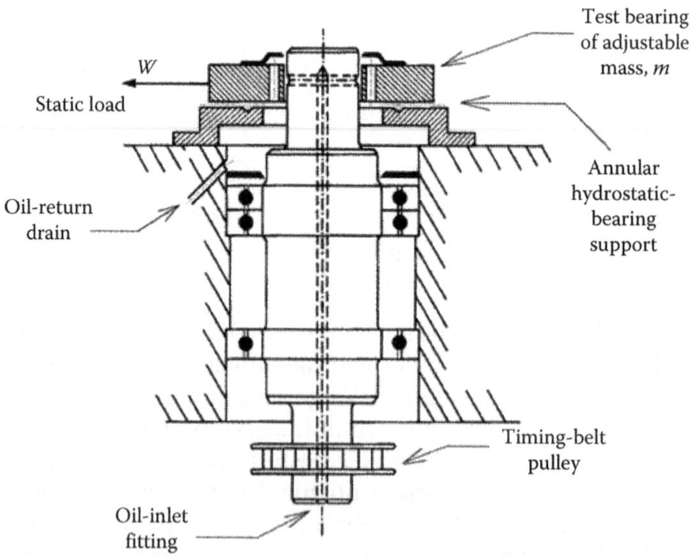

FIGURE 4.3
Test rig for controlled instability-threshold-speed research.

interactive dynamic radial force vector $\{P\}$ acting upon the rotor at a journal bearing is embodied only in the symmetric part "s" of $[c_{ij}]$ and the skew-symmetric part "ss" of $[k_{ij}]$, expressible as follows:

$$\begin{Bmatrix} P_x \\ P_y \end{Bmatrix} = -\begin{bmatrix} c_{xx}^{s} & c_{xy}^{s} \\ c_{xy}^{s} & c_{yy}^{s} \end{bmatrix}\begin{Bmatrix} \dot{x} \\ \dot{y} \end{Bmatrix} - \begin{bmatrix} 0 & k_{xy}^{ss} \\ -k_{xy}^{ss} & 0 \end{bmatrix}\begin{Bmatrix} x \\ y \end{Bmatrix} \tag{4.4}$$

The parametric equations $x = X\sin(\Omega t + \varphi_x)$ with $y = Y\sin(\Omega t + \varphi_y)$ are used here to specify a harmonic rotor orbit for the purpose of formulating the energy imparted to the rotor per cycle of a harmonic rotor displacement orbit, as follows:

$$E_{cyc} = \oint \left(P_x dx + P_y dy\right) = \int_0^{2\pi/\Omega} \left(P_x \dot{x}\, dt + P_y \dot{y}\, dt\right); \quad \text{i.e., } dx = \frac{dx}{dt}dt \text{ and } dy = \frac{dy}{dt}dt$$

$$= -\pi\left\{\Omega\left[c_{xx}^{s}X^2 + 2c_{xy}^{s}XY\cos\left(\phi_x - \phi_y\right) + c_{yy}^{s}Y^2\right] - 2k_{xy}^{ss}XY\sin\left(\phi_x - \phi_y\right)\right\} \tag{4.5}$$

By casting in the x–y system orientation of the principal "p" coordinates of $[c_{ij}^{s}]$, the c_{xy}^{s} term in Equation 4.5 disappears, yielding the following result, which is optimum for an explanation of rotor dynamic instability self-excited vibration:

$$E_{cyc} = -\pi\left[\Omega\left(c_{xx}^{sp}X^2 + c_{yy}^{sp}Y^2\right) - 2k_{xy}^{ss}XY\sin\left(\phi_x - \phi_y\right)\right] \tag{4.6}$$

Since $[k_{ij}^{ss}]$ is an isotropic tensor, its coefficients are invariant under orthogonal transformation, that is, do not change in transformation into the principal coordinates of $[c_{ij}^{s}]$. As detailed by the author (1987, 2010), realizing that exactly at the instability threshold speed $E_{cyc} = 0$, Equation 4.6 then provides proof for the following universally observed observation:

> The oil whip initiated rotor dynamic self-excited vibration is generally the lowest co-rotational (forward-whirling) rotor-bearing natural frequency mode which becomes zero damped when the instability threshold is reached.

Through the adjustment of the test bearing mass m, by adding or subtracting weights, one can vary the instability-threshold natural frequency of this 2-DOF system and thereby cause an instability threshold at selected operating conditions spanning a wide range of journal bearing Sommerfeld number (bearing dimensionless speed). The controlled test parameters are as follows:

1. Rotational speed ω
2. Bearing static radial load W
3. Lubricant viscosity μ
4. Test bearing mass m

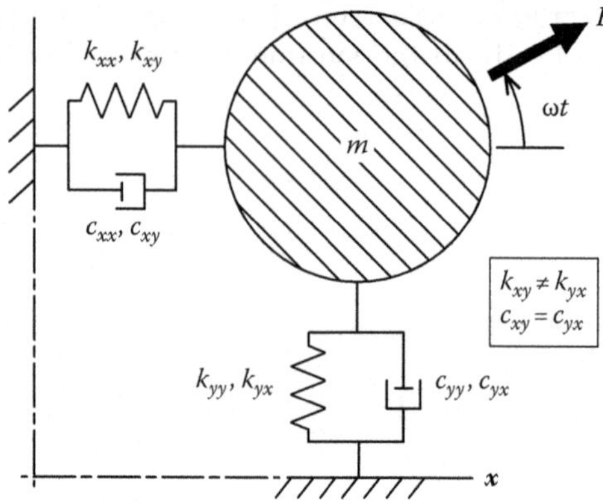

FIGURE 4.4
2-DOF model basis for Equation 4.7 ($F_0 = 0$).

Consistent with mechanical impedance approaches, the test measurements are correlated with an equivalent 2-DOF model given by the following equations and illustrated in Figure 4.4:

$$\begin{aligned} &m\ddot{x} + c_{xx}\dot{x} + k_{xx}x + c_{xy}\dot{y} + k_{xy}y = 0 \\ &m\ddot{y} + c_{yy}\dot{y} + k_{yy}y + c_{yx}\dot{x} + k_{yx}x = 0 \quad \text{with } c_{xy} = c_{yx} \end{aligned} \tag{4.7}$$

The complete procedure for extracting journal bearing dynamic coefficients at a given Sommerfeld number is summarized by the following sequence of steps:

1. Determine the stiffness coefficients using experimental or computed data.
2. Slowly increase spin speed until instability threshold speed is reached.
3. Capture x–y signals of "linear" instability growth; see Figure 4.5(a).
4. Invert the eigenvalue problem of Equation 4.7 to solve for the damping coefficients.

Basically, this procedure yields a *matched set* of journal bearing stiffness and damping coefficients. Even if the individual stiffness coefficients are from experimental static-load measurements, they are "matched" to the damping coefficients to replicate the experimentally observed and quantified instability threshold. Step 4 uses the experimentally observed instability threshold *frequency* and *orbit* parameters for the 2-DOF model in Equation 4.7. The Step-4 algorithm also uses the following information as inputs for the iterative eigenvalue inversion algorithm (Adams and Rashidi 1985).

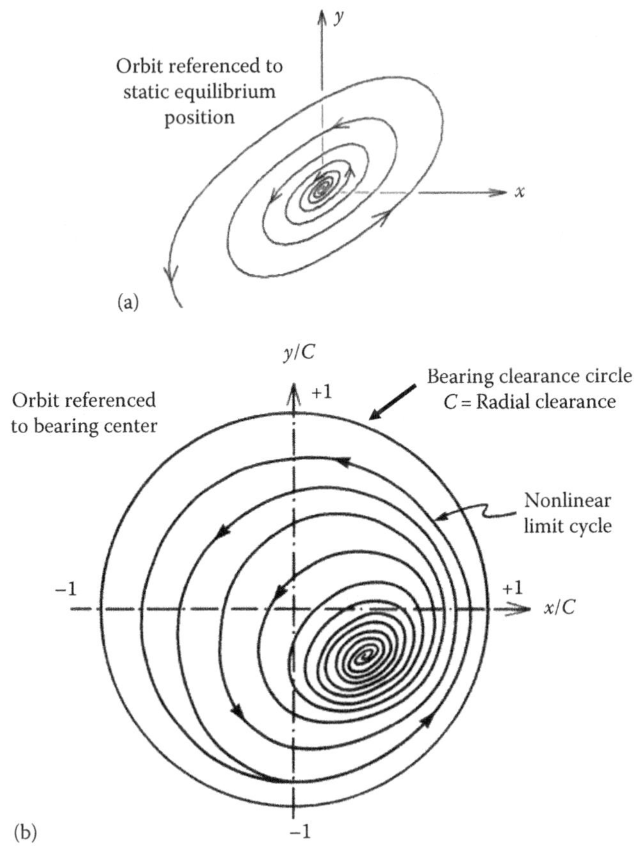

FIGURE 4.5
Transient orbital vibration buildup in an unstable condition: (a) initial linear transient buildup, and (b) growth to nonlinear limit cycle.

Data input for stability-threshold-matched bearing coefficients

1. Test bearing mass
2. Stiffness coefficients
3. Measured instability threshold speed, ω_{th}
4. Measured self-excited orbital vibration frequency Ω at instability threshold
5. x and y single-peak amplitudes ratio X/Y at the instability threshold
6. Phase angles' difference between x and y harmonic signals $\Delta\theta_{xy} \equiv \theta_x - \theta_y$

Using a standard eigensolution algorithm, the iterative computation determines the damping coefficient values (c_{xx}, $c_{xy} = c_{yx}$, c_{yy}) that, in combination with a priori stiffness coefficient values, reproduce the experimentally observed dynamic instability threshold information, that is, the observed (1) self-excited vibration threshold speed ω_{th}, (2) vibration frequency Ω, (3) normalized orbit ellipse shape (X/Y), and (4) angular orientation (X/Y) $\Delta\theta_{xy} \equiv \theta_x - \theta_y$.

Using postulated experimental measurement errors (Rashidi and Adams 1988) conclusively show the inherent superiority of this approach over mechanical impedance approaches, that is, the provision of a drastically improved accuracy for predicting not only (1) instability threshold speed but also (2) rotor vibration peak amplitude of an unbalance-excited critical-speed resonance. Clearly, the inherent sensitivity of the instability threshold phenomenon to damping provides the most accurate phenomenon from which to measure journal bearing damping parameters.

Equations 4.7 shows that in contrast to mechanical impedance approaches (Sections 4.1 and 4.2), dynamic force measurements are not needed in this approach, thus also eliminating a source of measurement error. However, the fundamental superiority of this approach lies in its basis that the "matched" stiffness and damping coefficients are consistent with $E_{cyc} = 0$ when the steady operating condition is nursed to its instability threshold condition.

4.4 Test Rig, High-Pressure Gas Labyrinth Seal Dynamic Forces

The size of the power plant's main steam turbine generator (TG) units increased considerably in the post-WWII period and well into the 1960s, from 100 MW to over 800 MW power ratings for single drivelines. Correspondingly, the turbine steam injection pressures and overall physical sizes of these gigantic machines also increased considerably. As always happens in any such drastic machinery evolution, a variety of new types of operating problems will show up in the field in spite of all the engineering development care employed along the evolutionary process. Addressing and solving those "new problems" through intense investigations can be justifiably characterized as "that's progress." One of those "new problems" was a large amplitude subsynchronous rotor vibration in the high-pressure turbine end of those multi-turbine TG rigidly coupled drivelines. It manifested itself quite similarly to journal bearing oil whip, discovered and identified in the early 1920s (Newkirk and Taylor 1925). But it was a distinctly different source for the encountered self-excited subsynchronous rotor vibration, eventually named "steam-whirl." Unlike oil whip, which is encountered as the rotor speed transgresses over the instability threshold speed before reaching operating speed, steam-whirl is encountered as the power output is being ramped up and occurs at a power output short of the unit's full rated power capacity. In terms of operating availability, *oil whip is worse than steam-whirl*, because the TG cannot be taken up to operating speed until the oil whip is extinguished, for example, by trial-and-error TG vertical bearing alignment adjustments. On the other hand, at least when steam-whirl occurs at a unit power output load less than the TG's rated full capacity, the unit can still generate power, albeit at power loads below the steam-whirl onset load, while the pending resolution of the steam-whirl problem is in process.

The author (2010) documents two power plant case studies where he successfully diagnosed and supplied the solution for eliminating steam-whirl problems, thus then allowing the TG units to operate at their full rated capacities. A rotor vibration phenomenon similar to steam whirl was also discovered during the same post–WW-II time period during the evolution of modern gas-turbine aircraft jet engines (Alford 1965). Similarly, 40 years earlier following WW-I, the previous evolution in steam turbine generator rotor speeds from 1800 rpm (4-pole at 60 Hz) to 3600 rpm (4-pole at 60 Hz), the oil whip phenomenon was discovered and identified (Newkirk and Taylor 1925). Even earlier (1890s), Gustav De Laval courageously proved by showing on an actual turbine that steam turbines could be safely operated at speeds exceeding their critical speed, leading the way for turbines to drive AC generators. That is 1800 rpm for 60 Hz 4-pole generators and eventually 3600 rpm for 60 Hz 2-pole generators.

Figure 4.6 illustrates the first explanation of this self-excited rotor vibration phenomenon, frequently referred to as the "Alford effect" in the United States (Alford 1965), on jet engines, and in Germany referred to as the "Thomas effect" (Thomas 1958), on steam turbines. When laboratory measurements of these labyrinth-seal-entrance effects were utilized in attempts to improve rotor dynamic analysis design predictions, it became apparent that the Thomas–Alford phenomenon only captured at most half of the full strength of the steam-whirl excitation force. That led to further analyses pointing to a complementary phenomenon, the flow field within the annular cavities between the labyrinth seal teeth. That led to significant proprietary research projects both in the United States and abroad. The author became more aware of the related research abroad during his 1984 summer consulting stay at the Large Steam Turbine Division of Brown-Boveri Corporation (BBC) in Baden, Switzerland, near Zurich. At BBC, resulting from the then recent prior paper on the subject, Adams (1980), the author, spent the 1984 summer showing BBC engineers how to analyze the very large amplitude rotor vibration that immediately ensues following a while-operating-detachment of large last-stage low-pressure steam turbine blades. During that time period, the author also visited Professor H. J. Thomas at Technical University, Munich to see his steam-whirl-related research test rigs. The Westinghouse Corporation (Pittsburgh) R & D Center's steam-whirl research test rig is described in this

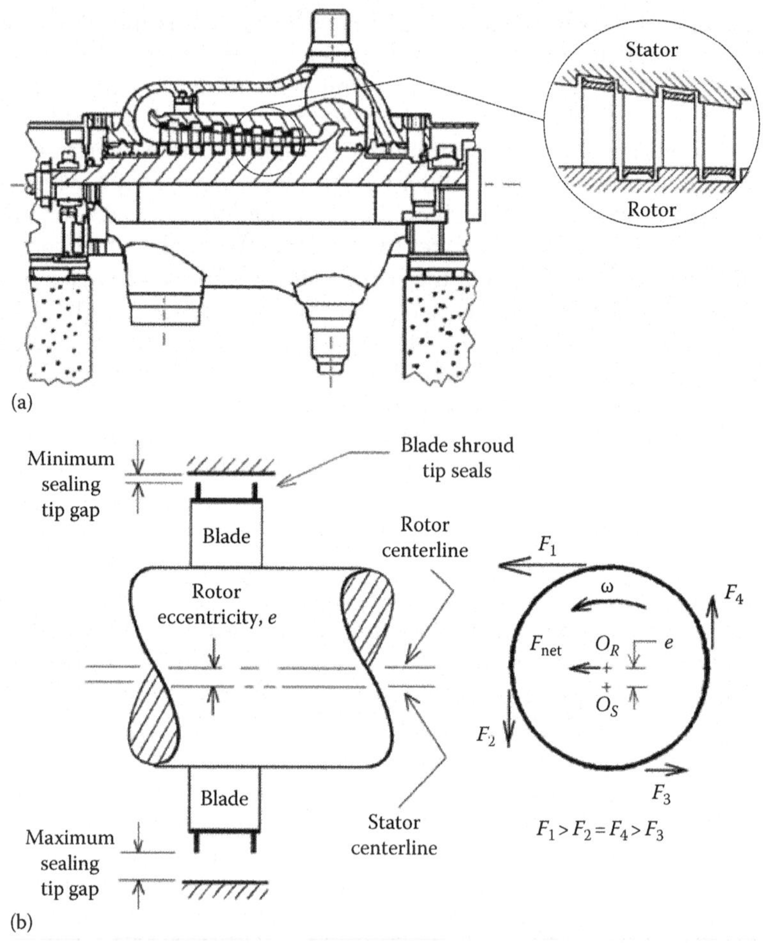

FIGURE 4.6
Contribution to *steam whirl* from the Thomas–Alford phenomenon. (a) Sectional view of a single-flow high-pressure steam turbine and (b) nonuniform torque distribution resulting from eccentricity.

section (Wright 1983), since it is probably the best and most impressive example detailed in the open literature.

The quite elaborate Westinghouse rig, illustrated by D. V. "Kirk" Wright (1983), is illustrated here in Figure 4.7. A fine attention to detail is provided by Wright. Having collaborated frequently with Kirk Wright in the 1970s, his thorough attention to details displayed in Figure 4.7 does not surprise the author one bit. Also presenting a paper at the same symposium, MIT Professor Stephen H. Crandall (1983) in the Q & A following Wright's presentation commented that Wright's experimental setup and project was perhaps the most impressive he had ever seen in its specialty.

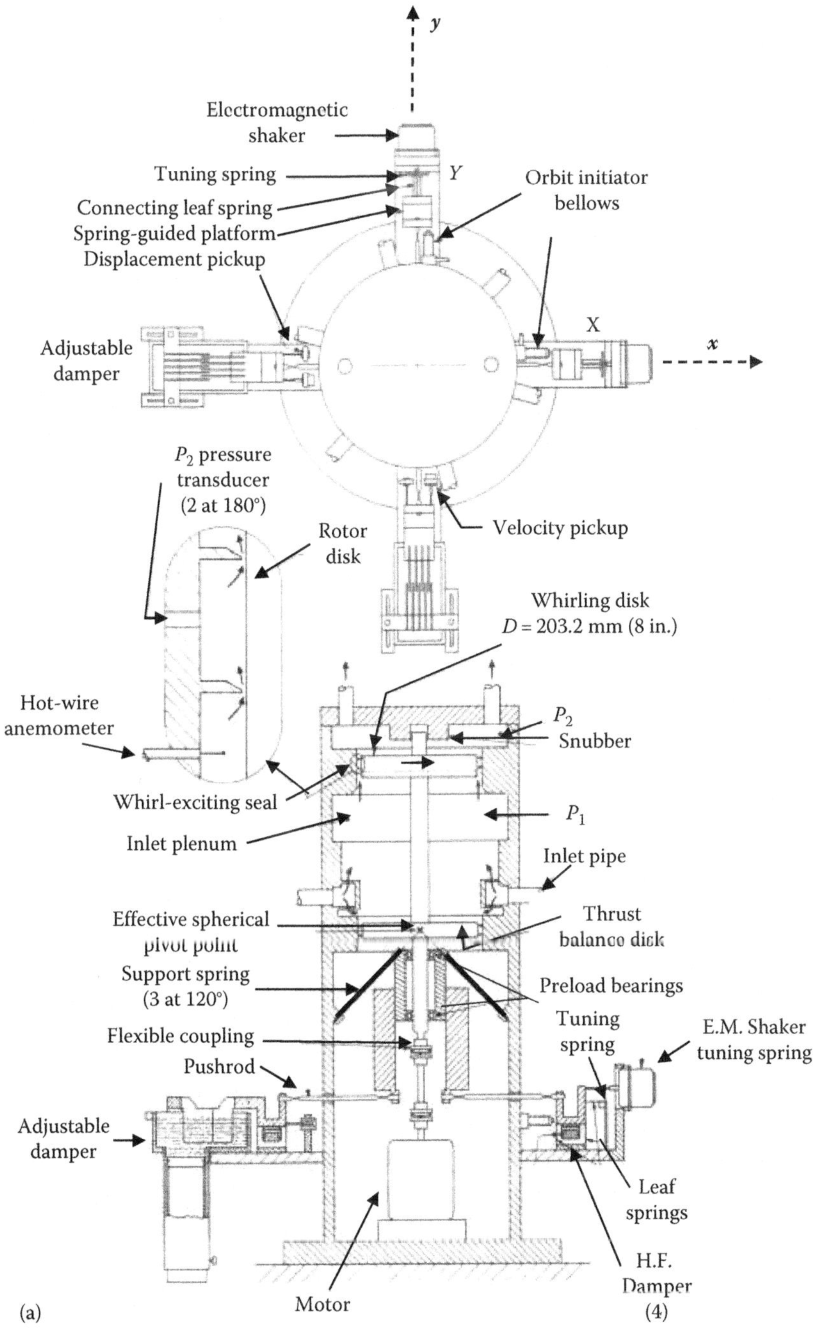

FIGURE 4.7
Labyrinth radial gas seal rig to measure rotor dynamic forces. (a) Seal-excited rotor whirl rig.
(Continued)

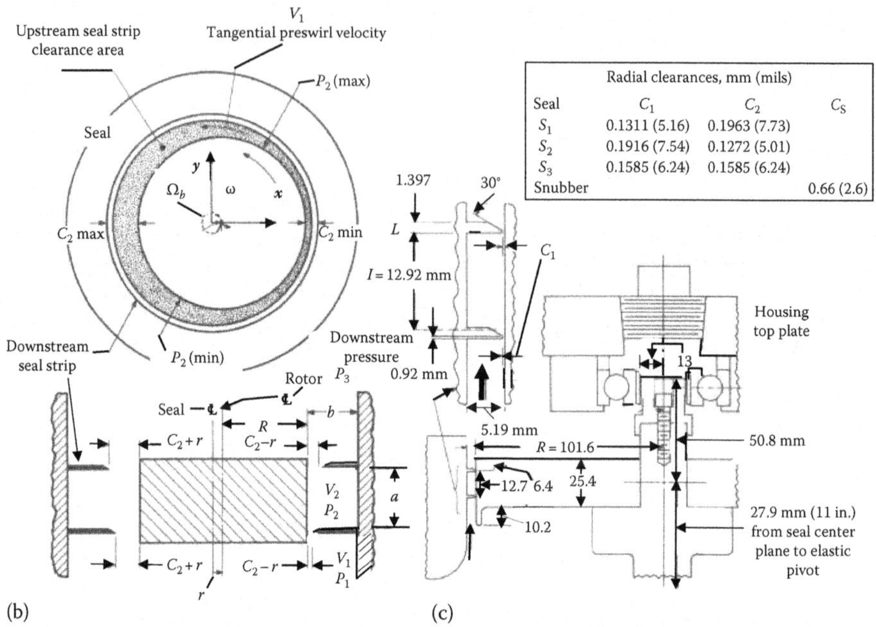

FIGURE 4.7 (*Continued*)
Labyrinth radial gas seal rig to measure rotor dynamic forces. (b) Diverging-seal forces on a backward-whirling rotor; (c) seal, disk, and snubber dimensions.

Wright's paper provides all the details on how the rig functions and a wealth of experimental results. Wright's 10 *Conclusions* are extensive and detailed. Here summarized are a few major conclusions:

1. The measure air labyrinth seal forces on the subsynchronously whirling model rotor, which are given in the paper, are accurate to about 3%. The experimental results can aid development of a valid analytical seal force prediction method that can be used in the design of turbines, centrifugal pumps, and compressors to avoid self-excited rotor whirl.
2. It should not be assumed that the model seal behavior directly predicts the behavior of large high-pressure turbine seals, which may have quite different characteristics because of their much larger dimensions, rotor surface speed, pre-swirl velocity, gas density, and rotor friction factor. However, a theory that accurately predicts the behavior of the model should also be valid for large seals.
3. The diverging model seal is strongly destabilizing for backward whirl. The converging model is stabilizing for both forward and backward whirls. Again, it is emphasized that these model results

are not to be generalized to larger seals only for use to validate any general theory analysis algorithm that could then be used with confidence to predict rotor dynamic behavior of large seals.

Bibliography

Adams, M. L., Non-linear dynamics of flexible multi-bearing rotors, *Journal of Sound and Vibration*, 71 (1), 129–144, 1980.

Adams, M. L., Insights into linearized rotor dynamics, Part 2, *Journal of Sound & Vibration*, 112 (1), 97–100, 1987.

Adams, M. L., *ROTATING MACHINERY VIBRATION: From Analysis to Troubleshooting*, 2nd edn., Taylor & Francis, CRC Press, Baco Raton, FL, 465pp., 2010.

Adams, M. L. and Rashidi, M., On the use of rotor-bearing instability thresholds to accurately measure bearing rotordynamic properties, *ASME, Journal of Vibration, Stress and Reliability in Design*, 107 (4), 404–409, 1985.

Alford, J., Protecting turbomachinery from self-excited rotor whirl, *ASME Journal of Engineering for Power*, 87, 333–344, 1965.

Crandall, S. H., The physical nature of rotor instability mechanisms, *Symposium on Rotor Dynamical Instability*, New York, ASME Book, AMD-Vol. 55, Adams, M. L., ed., ASME Applied Mechanics Division, University of Houston, TX, pp. 1–18, 1983.

Marscher, W. D., The effect of fluid forces at various operating conditions on the vibrations of vertical turbine pumps, *Seminar by the Power Industries Division of IMechE*, London, England, Feb. 5, 1986.

Newkirk, B. L. and Taylor, H. D., Shaft whipping due to oil action in journal bearings, *General Electric Review*, 28, 559–568, 1925.

Nordmann, R. and Massmann, H., Identification of stiffness, damping and mass coefficients for annular seals, *Proceedings of the Third IMechE International Conference on Vibration in Rotating Machinery*, York, England, pp. 167–181, 1984.

Rashidi, M. and Adams, M. L., Improvement to prediction accuracy of stability limits and resonance amplitudes using instability threshold-based journal bearing rotordynamic coefficients, *Fourth IMechE International Conference on, Vibrations in Rotating Machinery*, Heriot-Watt University, Edinburgh, Scotland, U.K., pp. 235–240, Sept. 1988.

Thomas, H. J., Instabile Eigenschwingungen Turbinenläufern angefacht, durch die Spaltstromungen Stopbushen und Beschauflungen [Unstable natural vibration of turbine rotors excited by the axial flow in stuffing boxes and blading], *Bull de L'AIM*, 71 (11/12), 1039–1063, 1958.

Wright, D. V., Labyrinth seal forces on a whirling rotor, *Symposium on Rotor Dynamical Instability*, New York, ASME Book, AMD-Vol. 55, Adams, M. L., ed., ASME Applied Mechanics Division, University of Houston, TX, pp. 19–31, 1983.

5

Model-Based Condition Monitoring of Nuclear Power Plant Pumps

5.1 New Multistage Centrifugal Pump Test Facility

The Case Western Reserve University (CWRU) multistage centrifugal pump test facility for research in *model-based conditioning monitoring* of power plant pumps is shown in Figure 5.1 (Michael Adams 2014–2016). Its development is relevant to centrifugal pump condition monitoring in general but was initially in response to specific power plant applications where the pumps are inaccessibly submerged and thus impractical for periodic condition inspections, for example, river pumps for nuclear power plants. Figure 5.2 shows the submerged accelerometer locations on the CWRU test pump inner casing surface. Figure 5.3 shows sensor locations on the test pump as it is submerged in the transparent test-loop outer can. A fairly new "player" in pump condition monitoring is the Robertson efficiency probes (Robertson and Baird 2015). A matched pair of these probes accurately measures pump efficiency in real time using ultraprecise temperature difference measurements, Figure 5.4. Trending pump *efficiency degradations* can be a valuable complementary indicator in assessing a pump's deteriorating health, thereby giving detection alerts that can help *avoid unscheduled outages.*

5.2 Model-Based Condition Monitoring

The fundamental premise of model-based condition monitoring is to reconstruct in real time the operating parameters, that is, *virtual sensors*, inside the pump from a computer model driven by measured signals from readily accessible external sensor locations, for example, on the pump driver unit. The term "virtual sensor" captures the underlying concept of the CWRU approach to model-based condition monitoring that is to reconstruct pump internal operating parameters, like shaft orbital vibration and internal

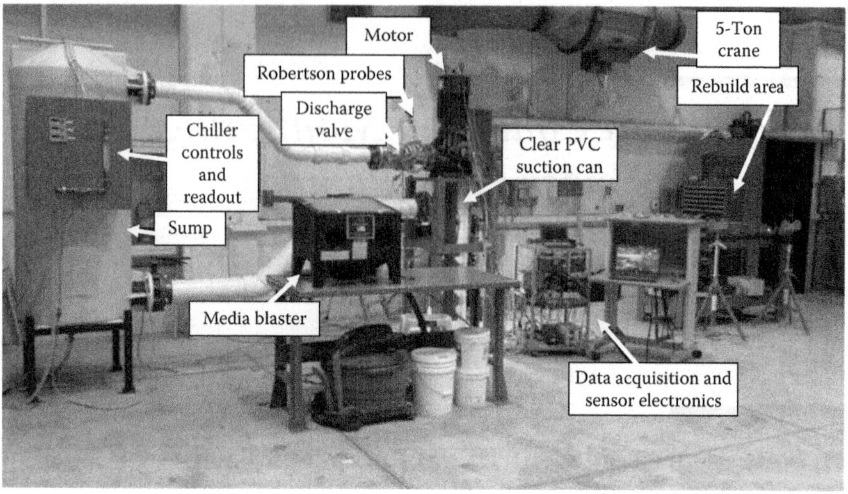

FIGURE 5.1
CWRU multistage centrifugal pump research test loop.

interstage seal leakage, where employing *real sensors* is not feasible. The CWRU test facility shown in Figures 5.1 through 5.4 is configured with sensors inside the unit, where it is not feasible to place real sensors in power-plant-operating pumps. And additionally, sensors are located at readily accessible external locations. In this research, competing real-time computer models of the pump are "tested" to determine their reliability in replicating the internal sensor measurements, driven only by the signals from the simultaneous external sensors. One could reasonably categorize this signal-replication process as an *advanced form of system identification* in which the interior behavior of the pump can be continuously monitored in real time. If proven successful, this research will advance the *condition monitoring* of many other types of power plant machinery, and likewise other industries' machinery, for example, chemical process plants, naval shipborne propulsion drivelines, water supply systems, and others.

If future major applications of rotating machinery are to be economically optimized in an environment of greatly *reduced maintenance personnel* and very few available *true experts* still around, then new yet-to-be-introduced *machinery management systems* will be required. The development of such new systems is a long-standing topic of extensive ongoing research. For example, the CWRU team has developed model-based monitoring-prognostic software that incorporates an array of machine-specific vibration models, specific to an extensive array of operating modes as well as fault types and severity levels. In the flow chart shown in Figure 5.5 (Adams 2010), each computer model (called an "observer") operates in real time and encodes a particular operating mode or fault type and severity level, that is, each model that embodies a specific identifiable risk to the safe and continued operation of the machine.

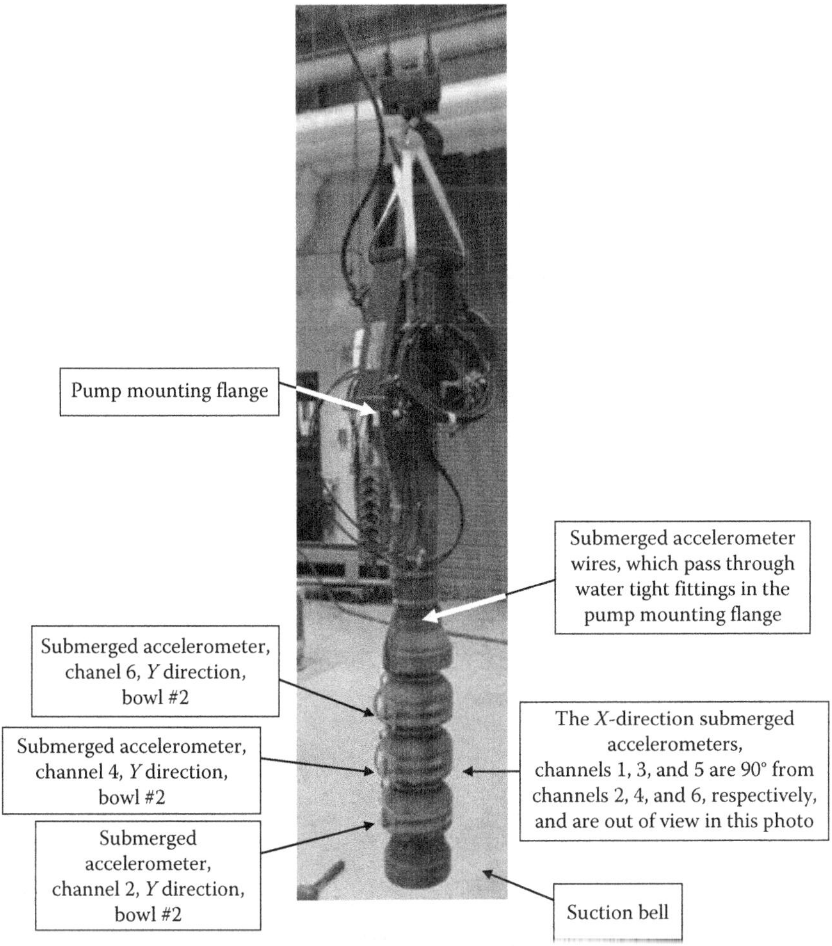

FIGURE 5.2
Submerged accelerometer locations on CWRU 3-stage test pump.

The outputs of these observers are the vibration signals that would be expected if the system was operating according to the model parameters of the observer. In this example, the observers are dynamic system models driven by the actual measured vibration signals from the machine being monitored. The observer outputs are continuously combined with the actual monitored vibration signals from the machine and correlated through a novel set of statistical algorithms and model-based filters, as summarized by Loparo and Adams (1998). In this real-time continuous process, each observer is driven by the monitored vibration signals of the machine and is being driven in a manner that makes each observer replicate or follow the actual monitored vibration signals as best as possible.

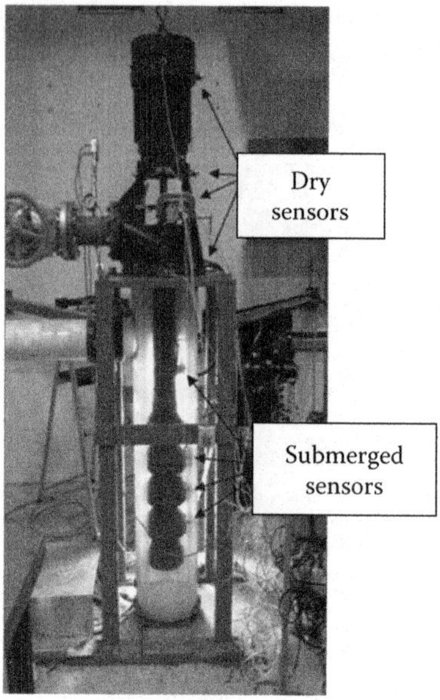

FIGURE 5.3
Sensor locations on installed CWRU 3-stage test pump.

FIGURE 5.4
Robertson pump-efficiency probes.

From the mismatch, that is, the residuals, between the measured vibration response of the machine and the predicted vibration response as produced by each observer, a set of probabilities are generated. These probabilities statistically quantify the match between each observer's output and the actual measured vibration signals for each fault type and severity level potentially

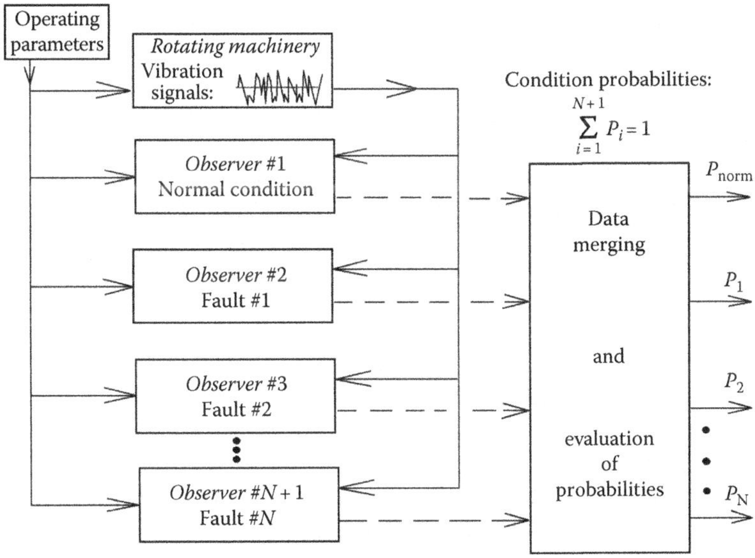

FIGURE 5.5
Real-time probabilities for defined faults and severity levels from the statistical correlation of monitored and model-predicted vibration signals.

in progress. The observer that best matches the current operating condition will have the highest probability in the set. That observer then identifies the operating problem(s) and severity level(s) most likely to be occurring at the current operating time. Note that this type of real-time monitoring system has the advantage of providing for simultaneous detection and isolation of faults and unwanted operating conditions. An important fact to realize when studying the example given in Figure 5.5 is the *ever-present vibration in any operating machine*; while excessive machinery vibration itself is generally considered to be an operating problem, even acceptable moderate levels of vibration have been conclusively shown to contain a wide range of informative diagnostic information for many of the other sources of a machine's deteriorating health. Vital life signs of a machine can include its vibration, acoustic noise, internal pressures, temperatures, performance indicators like efficiency, and radiation emissions.

The dynamic models of the machine's vibration response encoded in the observers illustrated in Figure 5.5 are also employed to remove sensor measurement signal "noise" that does not statistically correlate with the model's response. In contrast to conventional signal-noise-filtering techniques where signal noise is removed based on time-signal morphology or frequency domain characteristics, such model-based statistical-correlation filtering allows the retention of low-level and fine-structure signal components that are correlated with the physical model. One of the many interesting and important findings by the CWRU team is that the various fault and fault-level-specific observer vibration models do not have to be as "nearly perfect" as one might suspect, because the total of all observer probabilities is constrained to a *sum of one*. So a model (observer) need only be representative enough of its respective operating mode to *win the probability race* among all the other observers when its fault or fault combination type(s) and severity level(s) are the dominant operating conditions. Compared to the rule-based approach inherent in so-called *expert systems*, this *physical model-based* statistical approach is fundamentally much more open to correct and early diagnoses. And it is fundamentally more likely to identify infrequently encountered failure and maintenance-related phenomena, especially conditions not readily covered within a rule-based expert system.

An important additional benefit of a model-based diagnostics approach is the ability to combine measured vibration signals with observer vibration computer model outputs to make real-time determinations of rotor vibration signals at locations where no sensors are installed. This important factor is the genesis of the *virtual-sensor* research conducted in the new CWRU test facility shown in Figure 5.1. Typically in the field, vibration sensors are installed at or near the bearings, as dictated by sensor access to the rotor and sensor survivability. However, at midspan locations between the bearings is where operators would most like to measure vibration levels but cannot because of inaccessibility and the hostile environment for vibration sensors. Thus, the model-based approach provides *virtual sensors* at inaccessible rotor locations. Finally, another important attribute of the model-based approach is that it can readily be extended to include *prognostics* (Loparo and Adams 1998).

Since the observer models are derived from the physics of the machine, they can be used to provide a prediction of the equipment's future vibration response to additional faults, increased loading, etc. This predicted vibration response in conjunction with quantitative failure-mode analysis methods can be used to assess the remaining useful life (RUL) of the equipment. In this way, maintenance and operating personnel can be continuously aware not only of the current operating state of the equipment, they can additionally be informed of the equipment's likelihood to safely and reliably withstand future operating conditions that could be considered and potentially imposed, like operating at output power levels exceeding the machine's rated capacity.

It is also important to note that the model-based techniques proposed in the Figure 5.5 example are relevant not only to pumps as an individual unit but also to a motor–pump combined driveline. In the motor–pump system, the dynamics of the coupled motor–pump driveline, as well as the dynamic interactions between them, are included in the observer models. In this way, if it is determined that the motor is healthy, then information from the motor, such as operating voltages, phase currents, etc., can be employed as additional measured variables fed in real time into the observers.

The field of modern condition monitoring for rotating machinery is now well over 50 years into its development and thus is truly a mature technical subject. However, it continues to evolve and advance in response to new requirements to further reduce machinery downtime and drastically reduce maintenance costs. This driving motivation is particularly true for critical safety-related pumps in pressurized water reactor (PWR) and boiling water reactor (BWR) nuclear power plants.

5.3 Summary of Nuclear Power Plant Pumps

Cutaways of major pumps in a PWR nuclear power system are shown in Figure 5.6 for typical configurations now in service (Adams 2017). These illustrations are not all shown to the same scale. The Reactor Coolant Pump (RCP) is, of course, vitally important in maintaining constant normal heat removal from the reactor. A schematic of an entire PWR nuclear power plant is illustrated in Figure 5.7, showing where pumps are located in the overall plant configuration. Pump segments and specific components are listed in Table 5.1. The specific technology topic of *condition monitoring and prognostics* holds the key to optimum utilization of all the other important design technology areas listed in Figure 5.8 (Adams et al. 2004).

Dictated primarily by piping configurations, a number of major nuclear power plant pumps have vertical centerlines of rotation in contrast to most rotating machines, which typically have horizontal centerlines of rotation. That is because in the overall reactor system designs, the orientation of the pump inlet and discharge piping takes priority in determining whether the pumps have horizontal or vertical rotational centerlines. This is a fact both for commercial nuclear power plants as well as for naval nuclear-powered propulsion systems. An unintended and long-not-understood consequence of a vertical pump orientation was not discovered until the mid- to late 1970s, exposed by Makay and Adams (1979). Major examples of this include all PWR RCPs, all BWR primary reactor recirculating pumps, river pumps, and various other pumps.

The major property of these pumps with vertical centerlines that was not well understood until the late 1970s is the *high degree of static-indeterminacy of the radial bearing static loads*. Their radial bearings do not support the rotor weight, which for horizontal centerline rotors usually provide a major deadweight biasing of the radial-bearing loads. This means that for an accurate calculation of bearing static loads, both rotor and bearing static deflections must be factored into the bearing static load determinations.

Equally significant, without a deadweight biasing of radial-bearing loads, other static load sources often of secondary importance with horizontal centerline rotors become the major bearing static load contributors, for example,

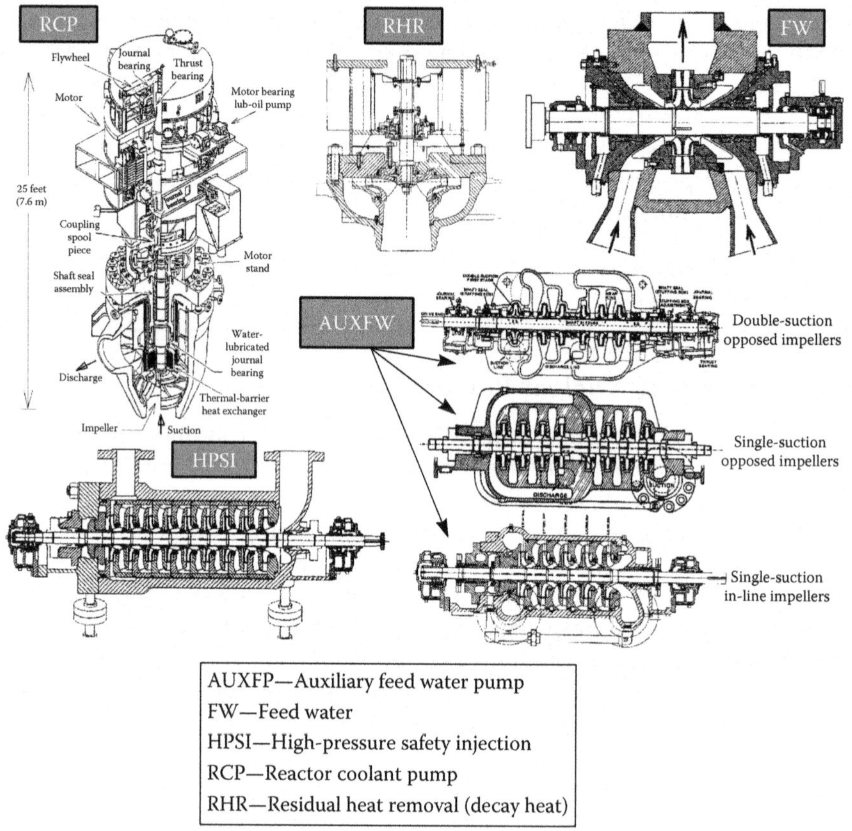

FIGURE 5.6
PWR nuclear power plant major safety-related centrifugal pumps.

motor electromagnetics, pump impeller forces, and other fluid–solid interaction effects like water-filled canned motor radial gaps (Adams and Padovan 1981). RCP configurations in most U.S. as well as several foreign nuclear power plants have employed the rigid-coupled 3-journal bearing rotor configuration typified by the RCP shown in Figure 5.6. That further accentuates the *high degree of indeterminacy of the radial-bearing static loads*.

Finally, this inherent property of vertical rotational centerlines comes with a high degree of additional uncertainty of radial-bearing loads resulting from radial manufacturing tolerances, assembly variations, pump operating flow points, and normal wear at close-clearance radial gaps (Adams 2010). In consequence, the *rotor vibration characteristics* of these pumps are also highly variable because of the radial bearings' strong dependency on bearing stiffness and damping properties upon bearing static radial loads. So it is likely that some troubleshooting "insight" gained from one machine may not be of value in assessing a vibration problem on another identically

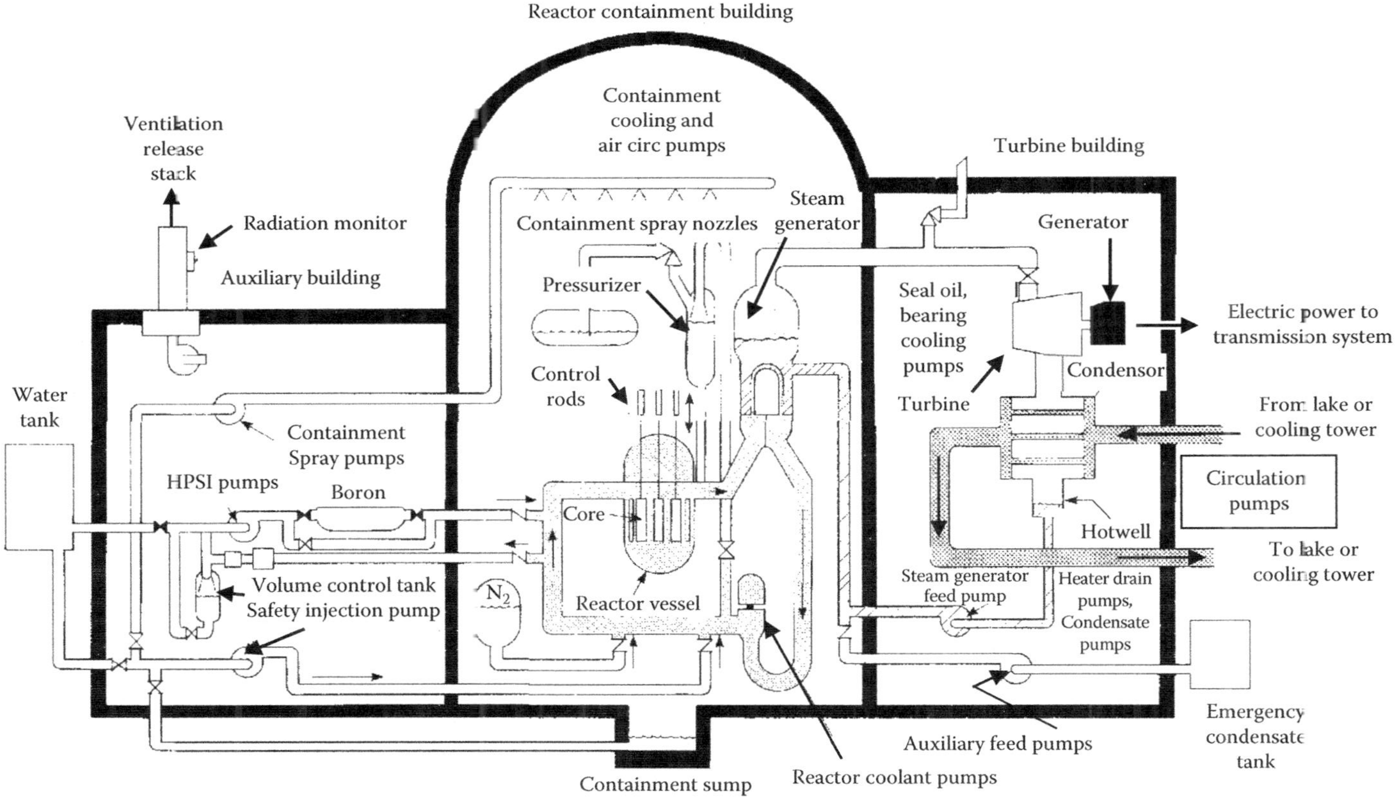

FIGURE 5.7
Essentials of a PWR nuclear power plant.

TABLE 5.1

Pump Segment and Component Summary List

Pump Segment	Parts
Rotating elements	Shaft, impellers, miscellaneous spacers' thrust runners, fasteners
Nonrotating internals	Diffusers or volutes, return channels, wear surfaces, fasteners
Pressure-containment casing	Upper casing, lower casing, fasteners, suction and discharge nozzles
Mechanical subsystems	Thrust bearing, radial bearings, shaft seals, wear rings' thrust balancer, coupling, fasteners,
Support	Base frame, fasteners

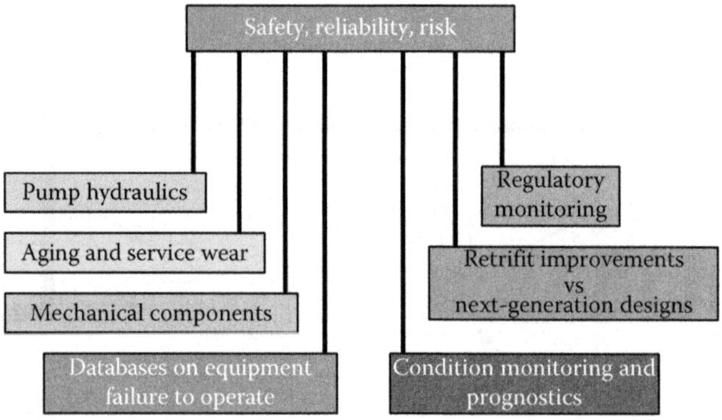

FIGURE 5.8
Important technology topics for PWR/BWR nuclear pump R & D.

designed pump. Even if a specific RCP is disassembled for inspection and/or repair, its vibration behavior may change significantly after reassembly. Jenkins (1993) attests to the considerable challenge in assessing the significance of monitored vibration signals from RCPs and thus focuses on possible correlation of vibration signal content and equipment malfunction as related to machine age.

Bibliography

Adams, M. L., *Rotating Machinery Vibration: From Analysis to Troubleshooting*, 2nd edn., Taylor & Francis, CRC Press, Boca Raton, FL, 465pp., 2010.

Adams, M. L., Model-based power plant centrifugal pump condition monitoring (Principal Investigator), EPRI Research Project, Charlotte, NC, 2014–2016.

Adams, M. L., *Power Plant Centrifugal Pumps: Problem Analysis and Trouble-Shooting*, Taylor & Francis, CRC Press, Boca Raton, FL, 182pp., 2017.

Adams, M. L., Loparo, K. A., Kadambi, J., and Zeng, D., Model-based condition monitoring for critical pumps in PWR and BWR nuclear power plants, Proposal to the U.S. Nuclear Regulatory Commission (NRC), Case Western Reserve University, Cleveland, OH, 63pp., Jan. 2004.

Adams, M. L. and Padovan, J., Insights into linearized rotor dynamics, *Journal of Sound and Vibration*, 76 (1), 129–142, 1981.

Jenkins, L. S., Troubleshooting Westinghouse reactor coolant pump vibrations, *EPRI Symposium on Trouble Shooting Power Plant Rotating Machinery Vibrations*, San Diego, CA, May 19–21, 1993.

Loparo, K. A. and Adams, M. L., Development of machinery monitoring diagnostic and prognostic methods based on nonlinear vibration characteristics, *Proceedings, 52nd Meeting of the Society for Machinery Failure Prevention*, Virginia Beach, VA, March 30–April 2, 1998.

Makay, E. and Adams, M. L., Operation and design evaluation of main coolant pumps for PWR and BWR service, Final Report, EPRI Project NP-1194, Charlotte, NC, 91pp., Sept. 1979.

Robertson, M. and Baird, A., Thermodynamic pump performance monitoring in power stations, *IMechE Seminar, Fluid Machinery in the Power Industry*, Bristol, U.K., June 10, 2015, www.robertson.technology.

[illegible]

6

Pumping Fluid-Solid-Particle Mixtures

Pump research at Case Western Reserve University (CWRU) includes the nuclear pump research test facility for model-based condition monitoring covered in Chapter 5 and also described by Adams (2017). The focus of this chapter is the CWRU pioneering experimental pump research headed by Professor Jaikrishnan Kadambi.

6.1 Particle Measurements in a Centrifugal Slurry Pump

Kadambi et al. (2007) measured the velocities of the slurry particles in the impeller of a centrifugal slurry pump using a Particle Image Velocimetry (PIV) technique in conjunction with refractive index matching. They conducted tests with an optically clear centrifugal slurry pump, Figure 6.1, at speeds of 725 and 1000 rpm using a slurry mixture comprising a sodium iodide fluid and 500 μm mean diameter glass beads. They found that the highest particle velocities occurred on the suction side of the impeller vanes and increased with pump speed but are less than the circumferential velocity of the vane tip. They also found that the average particle velocity decreases with an increase in particle concentration. Particle fluctuating velocity components from unsteady flow were also captured.

The facility for conducting these tests, Figure 6.2, is truly cutting-edge in fluid mechanics measurement technology. It employs an Nd-YaG laser, light-sheet-generating optics, optical encoder, and CCD camera and involves processing the conditioned measurement signals with PIVACQ and PIVPROC software to obtain particle velocity vectors extracted from two interrogation areas.

The relevance of this research is considerable since liquid-with-solid transport slurries are piped over short and moderate distances in many applications including *chemical and food processing*, *mining*, and *power generation*. Such slurry transport applications naturally incur the most severe erosive wear environment of all fluid transport systems. This results in high wear rates that limit the life of internal wetted component surfaces such as the pump impeller, wear rings, shaft seals, and casing. Wear to slurry pump internals understandably is a function of the (1) overall pump design; (2) operating conditions; (3) design specific speed; (4) operating flows as a percentage of the best-efficiency flow; (5) Net Positive Suction Head (NPSH), that is, propensity for cavitation; (6) internal component materials; and more (Adams 2017).

FIGURE 6.1
Centrifugal slurry pump with clear casing and clear impeller.

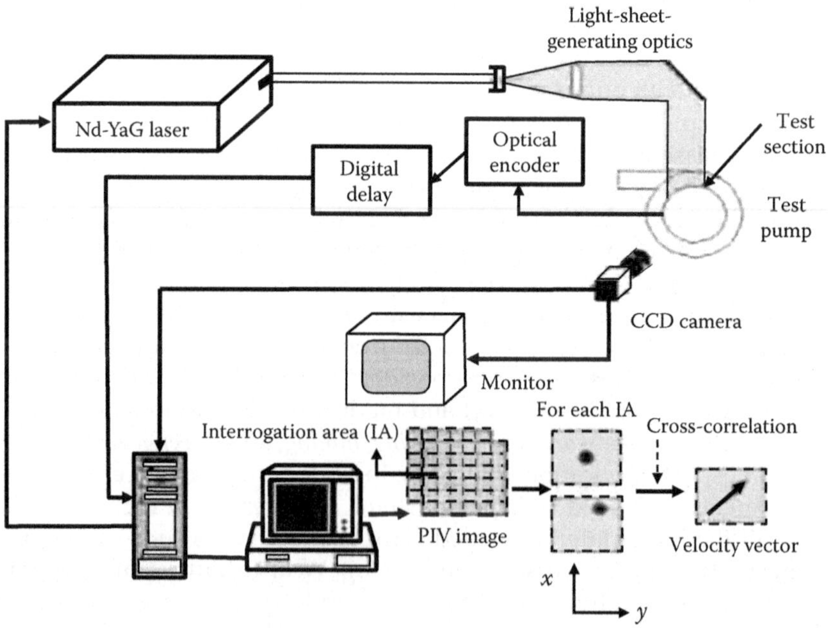

FIGURE 6.2
Schematic of CWRU centrifugal slurry pump PIV test rig.

Thus for example, altering pump internal component geometries provides avenues for extending their wear life. Clearly, nonintrusive methods to interrogate velocity flow fields in a pump eliminate the inherent compromises from flow-intrusive measurement techniques. The work by Kadambi and his team is thus an important contribution to the state of the art in slurry pump design technology.

6.2 Rotary Blood Pump

Professor Kadambi and his team constructed a transparent full-size rotary blood pump, Figure 6.3, to utilize PIV to measure its internal fluid velocities and turbulent flow stresses. The primary motivation for this work was to develop an understanding of the hemodynamics within blood flow that is critical to the development and validation of computational fluid dynamics (CFD) models for blood pump design analyses. They used a blood analog solution (BAS) of sodium iodide and glycerin to match the physiological kinematic viscosity of human blood. And the refractive indices of the BAS, pump casing, and impeller were matched to facilitate the use of PIV for fluid velocity measurements. The test rig for this work, Figure 6.4, was similar to that shown in Figure 6.2 for the earlier slurry pump research. Fluid velocity measurements made in the volute exit/diffuser region are presented for pumps speeds of 3000–3850 rpm (Sankovic et al. 2004), with the physiological pressure of 12 kPa under a maximum flow condition.

FIGURE 6.3
Transparent rotary blood pump.

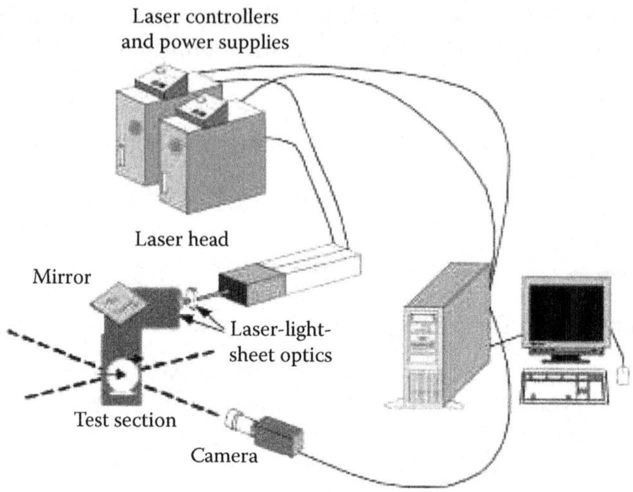

FIGURE 6.4
Schematic of a blood pump PIV test rig.

Hundreds of thousands of heart surgeries are performed in the United States yearly, including Professor Kadambi's own heart valve replacement in the 1990s. During surgery, the patient is placed on a blood pump along with an oxygenator. For artificial heart applications, blood pumping is extremely challenging. The pump must not cause hemolysis and thrombosis. Also, the reliability must be high, the power consumption low, and the unit size compact (Sankovic et al. 2004). A major advantage of a rotary pump over a pulsatile pump is the reduced physical pump volume, that is, one-tenth the volume of a pulsatile pump, allowing the insertion of the rotary pump inside the existing left ventricle. But the operating flow regime is at a much lower Reynolds number (10^3–10^4) than industrial pumps (2×10^4–10^8), dramatically increasing hydraulic losses. To further complicate matters, blood is a dense, high-particle-concentration, solid–liquid slurry, which exhibits non-Newtonian characteristics at low shear rates. Thus, to validate CFD models for blood pumps, the need for experimental flow field data is even greater than for industrial pumps.

Bibliography

Adams, M. L., *Power Plant Centrifugal Pumps: Problem Analysis and Trouble-Shooting*, Taylor & Francis, CRC Press, Boca Raton, FL, 182pp., 2017.

Furlan, J. M., Kadambi, J. R., Visintainer, R. J., and Garman, M., Local particle concentration measurements in a centrifugal slurry pump using a scan ultrasound technique, *14th International Symposium on Transport Phenomena and Dynamics of Rotating Machinery, ISROMAC-14m*, Honolulu, HI, 2012.

Kadambi, J. R., Mehta, M., Sastry, S., Sankovic, J., Wernet, M. P., Addie, G., and Visintainer, G. R., Particle velocities in the rotating impeller of a slurry pump, *Proceedings of the Joint ASME Symposium on Solid-Liquid Slurry Flow, ASME Fluids Engineering Summer Meeting*, San Diego, CA, July, 2007.

Mehta, M., Kadambi, J. R., Sastry, S., Sankovic, J. M., Wernet, M. P., and Addie, G., Particle velocities in the rotating impeller of a slurry pump, *Proceedings, Fifth Joint ASME/JSME Fluids Engineering Conference*, San Diego, CA, Aug. 2007.

Sankovic, J. M., Kadambi, J. R., Mehta, M., Smith, W. A., and Wernet, M. P., PIV investigations of the flow field in the volute of a rotary blood pump, *ASME Journal of Fluids Engineering*, 126, 730–734, 2004.

Sastry, S., Kadambi, J. R., Sankovic, J. M., and Izraelev, V., A particle image velocimetry study of flow in a bladeless rotary blood pump, *Sixth Pacific Symposium on Flow Visualization and Image Processing*, Honolulu, HI, 2007.

7

Ohio State Gas Turbine Lab

The Gas Turbine Laboratory (GTL) at Ohio State University (OSU) secured its international recognition when in 1995 Dr. Michael Dunn moved from Calspan in Buffalo to OSU in Columbus and became a professor in the highly respected OSU Engineering School. He brought with him from Calspan the key facilities of its gas turbine jet engine research laboratory and key personnel. He has since added new cutting-edge test facilities to the OSU GTL. Before moving his Calspan laboratory to Ohio State, Professor Dunn worked for 34 years in aviation research, including at Lockheed and Cornell Aero Lab, which became Calspan in 1978. He has over 200 published papers in his field. It was an exciting and intense experience for the author and his son, now Professor Michael Adams, when they were retained to collaborate with Professor Dunn and his OSU team on the 1999–2003 research project covered in Chapter 15. Under Professor Dunn, R & D projects at the OSU GTL have been prominently utilizing three major facilities: (1) a large shock tube, (2) an engine compressor in-ground full-speed-spin pit facility, and (3) a large spin pit facility. The OSU GTL facilities require many channels of very high-sampling rate digital data acquisition, Figure 7.1. The focus of this chapter is on the said OSU GTL facilities (1) and (3).

7.1 Shock Tube Test Facility

Shock tubes are used to replicate and direct blast waves at a tested target to study explosions and their consequences under controlled laboratory conditions. They are also prominently used to replicate aerodynamic flows under a wide range of temperatures and pressures that would be difficult to attain in other types of laboratory facilities. Shock tubes are also used to investigate compressible flow phenomena and gas phase combustion. They have recently been used in biomedical studies to determine how biological specimens are affected by blast waves. The first shock tube was constructed at Vieille, France in 1899. In the 1940s, shock tubes were employed in the United States, Canada, and Great Britain for research in mine safety, shock-induced explosions, and aerodynamics. In the 1950s, a shock tube was first employed at Caltech to study early applications of high-temperature chemical kinetics. In the 1960s, they were first used to study earth-orbital reentry

FIGURE 7.1
OSU GTL data acquisition system.

aeronautics. In the 1970s, shock tubes were employed at Stanford/NASA Ames with lasers to research chemical species detection. Shock tubes are now also used in astrophysics, intensive blast waves of nuclear detonation intensity, chemical kinetics, aerodynamics, radiation plus convection heat transfer, and spectroscopy.

A shock tube short-duration test starts with the high-pressure driver gas rapidly expanding as a result of a controlled rupturing of the primary diaphragm, creating a shock wave, Figure 7.2. The test gas is instantly compressed and thus heated to a high temperature by incident and reflected shock waves. Clearly, this is a sudden energy conversion phenomenon that transfers initially stored driver gas pressure energy to the test gas, yielding the high-temperature/high-pressure conditions in the test chamber where data is acquired by sensors. A wide range of temperatures and pressures are achievable, that is, 600–4000 K and 0.1–1000 atm.

The OSU shock tube in Professor Dunn's GTL is possibly the largest and most elaborate of its kind. It occupies a major building addition initially

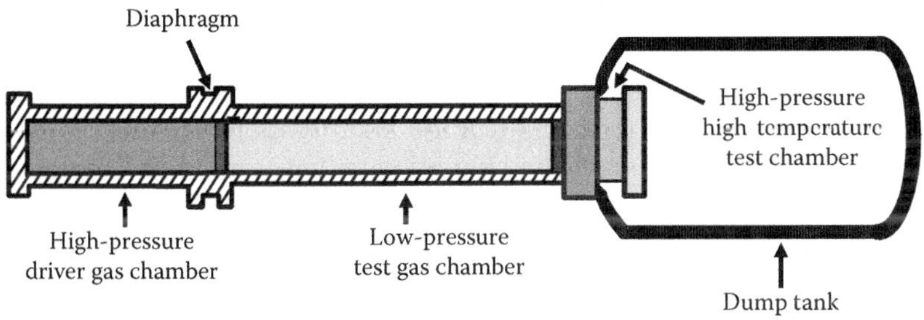

FIGURE 7.2
Schematic of a shock tube.

FIGURE 7.3
OSU GTL shock tube.

constructed just to house it, Figure 7.3. It can operate either in shock tube or blow down (see Section 17.2) mode. A primary mission of this shock tube facility has been to provide turbine designers an understanding of the fundamental flow physics involved and to accurately measure the needed empirical inputs to make accurate flow and thermal simulations with the modern Computational Fluid Dynamics (CFD) and Heat Transfer (HT) computer codes, now standard design analysis tools for gas turbines. Professor Dunn and his team have provided such vital information to nearly all of the gas turbine manufacturers. Quoting a recent historical review paper's opening statement (Dunn and Mathison 2013), "Short-duration facilities have been used for the past thirty five years to obtain measurements of heat transfer, aerodynamic loading, vibratory response, film-cooling influence, purge flow migration, and aero performance for full-stage high-pressure turbines operating at design corrected conditions of flow function, corrected speed, and stage pressure ratio."

Perhaps the most challenging of the many aircraft engine design advances continuing to evolve is increasing turbine inlet temperature and stage pressure ratio, with the well-established knowledge that as these parameters have been increased, substantial increases in performance efficiency have been achieved. The rise in turbine inlet temperature has been evolving for more than 50 years (Dunn and Mathison 2013). Short-duration shock tube tests have contributed significantly to this continuously evolving performance improvement. It will likely continue to evolve as the critically controlling technologies continue to advance. The fundamental facts that (1) turbine inlet flow is significantly unsteady and that (2) there is significant uncertainty for boundary conditions make the use of advanced CFD and HT computer codes heavily dependent upon the empirical correction factors extracted from shock tube facilities such as at OSU GTL.

TABLE 7.1

Professor Dunn's Shock Tube Turbine Projects, 1977–2012

1.	Garrett TFE731-2, 1977–2001
2.	Garrett LART, 1986
3.	Allison VBI, 1990–2000
4.	Teledyne 702, 1987–1989
5.	Rocketdyne SSME, 1994–1995
6.	P & W Vaneless counter rotating; 2000
7.	GE Vaneless counter rotating, 2000
8.	P & W, Early 6000, 2000–2003
9.	GE large single-stage turbine, 1997–2004
10.	Westinghouse ATS, 2003
11.	GE 1.5 Sage turbine (uncooled)
12.	Honeywell single-stage cooled turbine, 2006–2010
13.	GE 1.5 Sage turbine, 2010–2012
14.	Honeywell small turbine, 2012

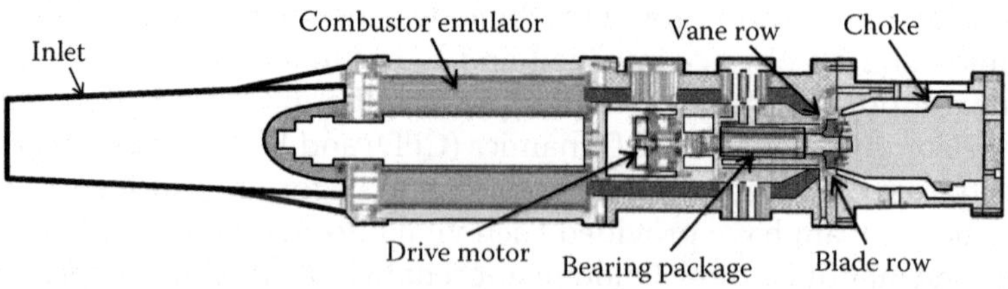

FIGURE 7.4
Shock tube turbine test chamber configuration. (From Haldeman, M. et al., Aero performance measurements for a fully cooled high-pressure turbine stage, *Proceedings, ASME Turbo Expo GT2012*, Copenhagen, Denmark, June 2012.)

In their paper, Dunn and Mathison provide a detailed treatment of how the shock tube facility now at the OSU GTL has been employed in several gas turbine research projects for many manufacturers' new engine developments. They tabulate those specific turbine development projects spanning the 1977–2012 period in more detail than briefly summarized here in Table 7.1. A detailed example of an OSU high-temperature, high-pressure turbine-stage shock-tube test chamber configuration is shown in Figure 7.4.

7.2 Large Spin Pit Facility

The OSU GTL's groundbreaking core-compressor blade-tip rub research for General Electric's Aircraft Engines Group in Evendale, OH, covered in Chapter 15, fostered the expansion of that research to study large blade-tip rub on engine fans and low-pressure turbines. To that end, a Large Spin Pit Facility (LSPF) was subsequently built at OSU GTL, first detailed in the open literature by Padova et al. (2011). The vertical spin pit stiff-spindle rotor for this facility, rated for 7000 rpm, is illustrated in Figure 7.5. An additional interchangeable stiff-spindle lower portion of the rotor, not shown here, is rated for 18,000 rpm. Causes of unavoidable blade-tip rub incursions in an aircraft engine during normal operation are detailed in Chapter 15.

The arrangement of the LSPF was specifically designed to investigate blade-tip rub phenomena of full-scale and scale models of contemporary fan and low-pressure turbine blades at representative engine speeds, with rotor diameters of up to 2.2 m. As explained in Chapter 15, blade-tip rub investigations dictate the use of a very high-stiffness spindle to virtually eliminate spindle deflections upon blade incursion with the shroud and to put rotor critical speed sufficiently above maximum spindle speed. As with the test rig covered in Chapter 15, full vacuum is maintained in the containment tank during operation to eliminate wind drag. Not shown in Figure 7.5 is an emergency brake.

Unlike the compressor blade-tip rub facility covered in Chapter 15, the blade incursion in the LSPF does not simulate the sudden tip rubs, such as from hard landings. The LSPF employs a slow progressive incursion mechanism, thus focusing more upon the blade-tip rub tribology and mechanics.

However, the LSPF casing rub sector and its three piezoelectric load measuring units (LMUs) are virtually a scaled-up version of the one covered in Chapter 15 (see Figure 15.5), here pictured in Figure 7.6 for the LSPF. Subsequent test results from the LSPF are presented by Langenbrunner et al. (2014). The cutaway of a typical modern high-bypass-flow fan jet engine is illustrated in Figure 7.7.

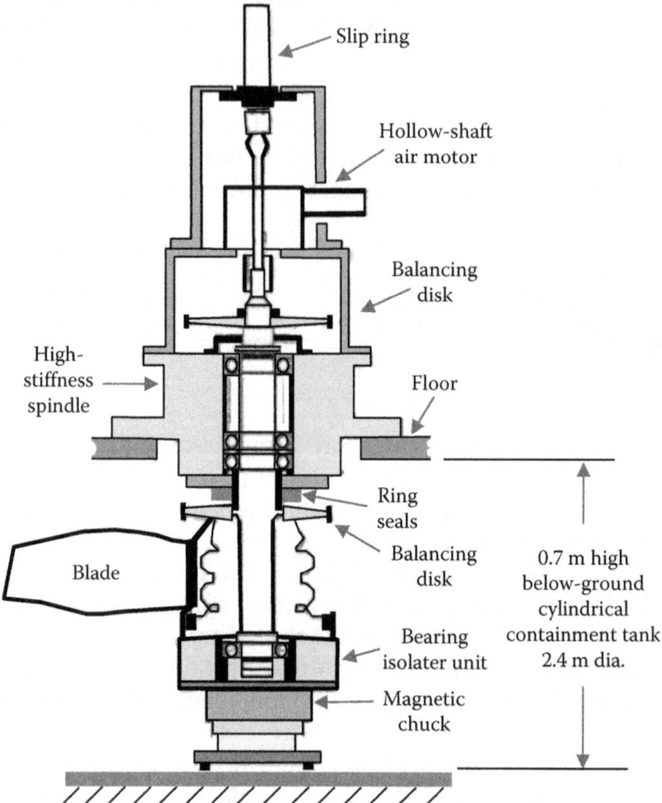

(a)

(b)

FIGURE 7.5
Section of LSPF rotor rated for 7000 rpm (incursion not shown): (a) sectional view of vertical spin fit rotor assembly and (b) photo of above floor portion of vertical spin fit rotor housing.

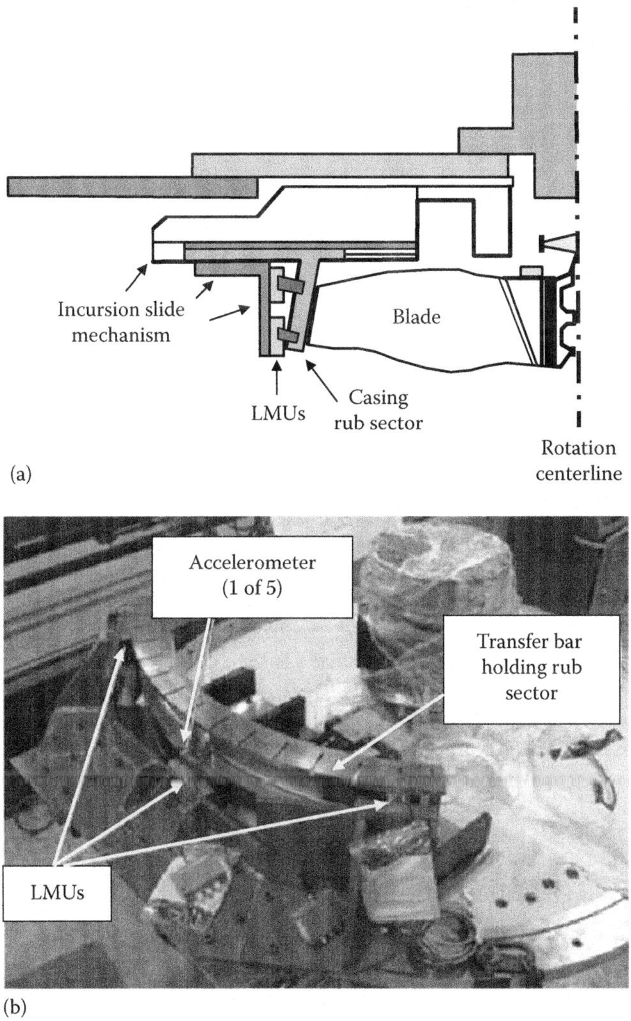

FIGURE 7.6
LSPF casing rub sector, incursion slide (a) illustration and (b) photo.

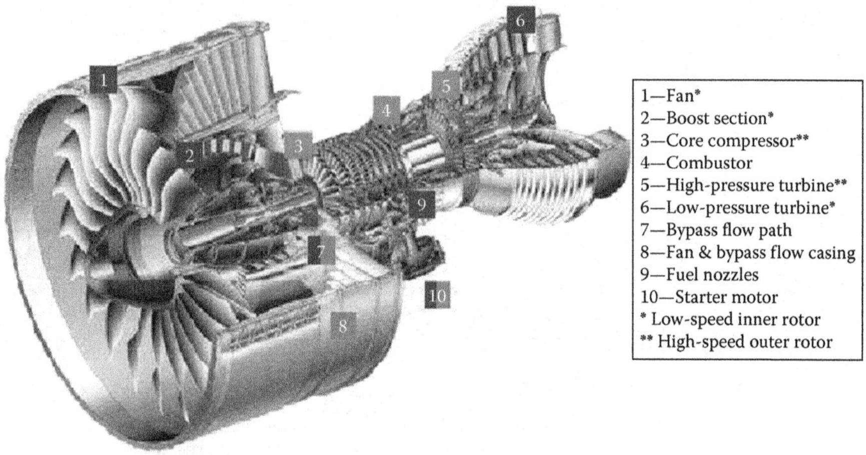

FIGURE 7.7
Modern high-bypass-flow fan jet engine.

Bibliography

Dunn, M. and Mathison, R., History of short-duration measurement programs related to gas turbine heat transfer, aerodynamics, and aero performance at Calspan and OSU, *Proceedings, ASME Turbo Expo GT2013*, San Antonio, TX, June 2013.

Haldeman, M., Dunn, M., Mathison, R., Troha, W., Vander Hoek, T., and Riahi, A., Aero performance measurements for a fully cooled high-pressure turbine stage, *Proceedings, ASME Turbo Expo GT2012*, Copenhagen, Denmark, June 2012.

Langenbrunner, N., Weaver, M., Dunn, M., Padova C., and Barton J., Dynamic response of a metal and a CMC turbine blade during a controlled rub event using a segmented shroud, *Proceedings, ASME Turbo Expo 2014*, Dusseldorf, Germany, June 2014.

Padova, C., Dunn, M., Barton, B., Turner, K., and Steen, T., Controlled fan blade tip/shroud rubs at engine conditions, *Proceedings, ASME Turbo Expo 2011*, Vancouver, British Columbia, Canada, June 2011.

8

Swiss Federal Institute Cavitation Research Facility at EPFL Lausanne

Cavitation in centrifugal pump impellers and hydro turbine runners is an ever-present important design and operation consideration. Operating these turbomachines with pronounced cavitation present can readily do serious erosive damage to pump impellers and turbine runners. In consequence, cavitation is one of the main controlling factors, which limit designing hydraulic machines with higher specific power and/or from setting them at higher elevations in the power plant. A comprehensive treatment of cavitation is given in the author's centrifugal pump book (2017), which describes the standard industry test method for determining a centrifugal pump stage's minimum required suction pressure (Net Positive Suction Head, NPSH) to avoid cavitation over its full operating flow range. The fundamental phenomenon of cavitation and the mechanism by which it causes serious erosive damage are quite complex. The basic understanding of cavitation was significantly advanced by the experimental research efforts undertaken in the Institute for Hydraulics and Fluid Mechanics at the Swiss Federal Institute École Polytechnique Fédérale de Lausanne (EPFL) in Switzerland. The underlying mission of the EPFL cavitation research was to provide improved insight into the cavitation physical process of industrial importance in the field of hydraulic machinery (Avellan et al. 1987).

Starting in the early 1960s, EPFL researchers conducted 20 years of continuous research using scale-model pump and turbine stages. In the early 1980s, with that accrued 20 years of experience, they sought to advance the field further by devising a new experimental facility not limited by the inherent difficulties of model testing with complex runner/impeller geometries and the difficulty of access to the internal flow field of a rotating runner/impeller. This chapter focuses on the elaborate experimental facility that was developed in this pioneering research leap to gain improved knowledge on the cavitation phenomenon as well as an expanded understanding of its accompanying erosion damage mechanism.

8.1 Description and Mission of Test Facility

The detailed layout and size of the EPFL facility is truly impressive, Figure 8.1. As part of an extensive $10M EPRI-sponsored research project on boiler feed water pumps, the author in 1987 visited this facility along with turbomachinery fluid mechanics specialist MIT Professor, David Gordon Wilson. The cost of this facility was then about SF 2M ($1M) in 1984 currencies. Development and construction of this facility was domestically sponsored by the Swiss Federal Government, the Swiss Energy Producers Association, Sulzer Pump Company (Winterthur), and the Vevey Engineering Works (Vevey).

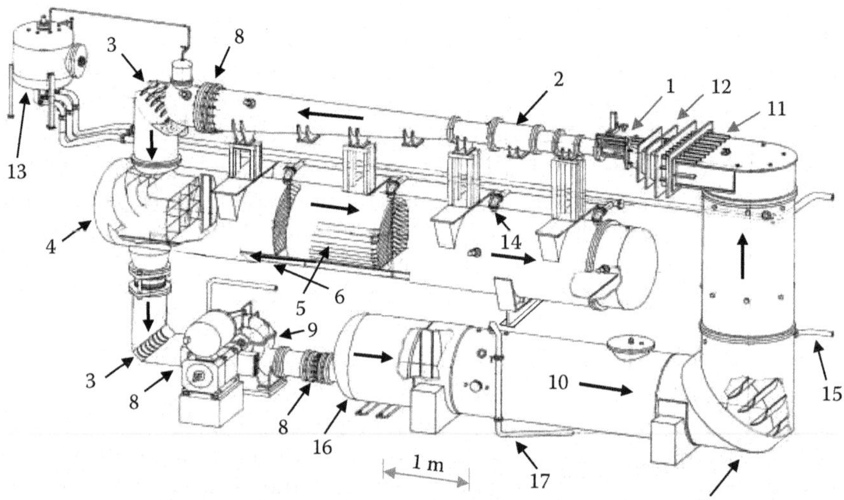

1—Test section
2—Diffuser
3—Vaned turn
4—Flow straightener elbow
5—Bubble trap
6—Return flow
7—Vaned turn
8—Expansion section
9—Circulating pump
10—Uniform flow developed
11—Settling chamber
12—Contraction nozzle
13—Pressure vessel
14—Bubble-collecting pipes
15—Cooling system pipes
16—Drains
17—High-pressure safety relief line
Flow direction →

FIGURE 8.1
EPFL high-speed cavitation tunnel. (From Avellan, F. et al., A new high-speed cavitation tunnel for cavitation studies in hydraulic machinery, *Proceedings, ASME Annual Meeting*, Boston, MA, December 1987.)

Major cavitation research areas of the EPFL with their new facility included the following three topics as detailed by Avellan and Henry (1984), Favre and Walther (1986), Karimi and Avellan (1986), and Avellan and Henry (1987):

1. Similarity dimensionless numbers relating model tests to full-size prototypes
2. Improved cavitation performance via the inverse-impeller modification method
3. Cavitation erosion damage rate prediction in industrial hydraulic machines

The new facility was thereby justified by promising the following deliverables:

1. Scientific basis for cavitation influence on hydraulic machine performance
2. Deep insight into the fine-structure flow field cause of cavitation erosion
3. Experimental database for hydraulic machinery cavitation erosion prediction

In preparation for the development of this new cavitation test facility, preliminary studies were conducted in the 1981–1982 time period. These included an extensive literature search, and EPFL researchers visiting many other hydraulics laboratories, for example, Sulzer Pump Co. in Winterthur; Institut de Mechique de Grenoble, Neyrtec, in France; ARL at the Pennsylvania State University; Caltech in Pasadena; and the St. Anthony Falls Hydraulic Laboratory in the United States. At the same time, EPFL researchers conducted hydraulic design and bubble-trap testing to complete the specification of the tunnel and its associated instrumentation. During 1983, a Perspex 1/10 scale model of the tunnel was constructed, Figure 8.2, to optimize design of the vaned turns and the bubble-trap straightener. The first operation of the completed facility started in March 1984.

8.2 Functioning of Cavitation Tunnel

Considering the scientific objectives of the new facility, considerable attention was devoted to instrumentation specifications. All macroscopic flow parameters had to be measured and kept constant. These include the (1) cavitation number, (2) fluid velocity, (3) cavitation nuclei content, and (4) dissolved gas content, all at the test section hydrofoil inlet. Furthermore, all that information needed to be displayed and recorded in real time.

FIGURE 8.2
1/10 scale model of the tunnel flow circuit. (From Avellan, F. et al., A new high-speed cavitation tunnel for cavitation studies in hydraulic machinery, *Proceedings, ASME Annual Meeting*, Boston, MA, December 1987.)

Each test specimen was a two-dimensional hydrofoil comparable to actual impeller vane geometries. Special instrumentation was needed to measure the cavitation flow parameters around the test airfoil. Those signals included (1) mean and fluctuating two-dimensional velocity components, (2) radiated cavitation noise, and (3) lift and drag fluid forces imposed upon the test hydrofoil. Given the high fluid flow velocity plus the cavitation phenomenon, it was required to employ a 2-component laser-doppler anemometer. Measurement of the other parameters did not require any new features in the remaining instrumentation, that is, (1) acoustic measurements and (2) hydrofoil force measurement restraints. The transit time for a complete fluid particle trip around the tunnel loop was 98 seconds at the maximum flow rate.

8.3 Early Insights Gained

As prior early research uncovered, centrifugal pump cavitation starts with the formation of vapor pockets or cavities (steam bubbles) in any flowing liquid when the liquid flows into a region where the pressure becomes lower than the liquid's vapor pressure. Of course, this fundamental phenomenon is not restricted to pumps and turbines.

In centrifugal pumps, cavitation is most likely to occur at the inlet (suction) region of the impeller on the low-pressure (trailing) side of the impeller vanes. In hydro turbines, cavitation is most likely to occur at the runner discharge region on the low-pressure side of the runner blades. For pump impellers, as these vapor pockets are swept with the flow further into the impeller, they experience the progressive increase in liquid pressure that the pump impeller naturally produces. These vapor pockets will therefore collapse inside the impeller. But because the transient from vapor back into liquid involves thermodynamic heat transfer, the collapse of the vapor pockets does not occur immediately upon experiencing a pressure that just exceeds the vapor pressure. The very small but finite thermodynamic time delay for the vapor pockets to collapse means that the collapse will occur at points where the local pressure has already significantly exceeded the vapor pressure. In consequence, the collapse of the vapor pockets occurs as violent implosions, which act to progressively erode the impeller vane surface when the pump is operated with *insufficient suction pressure* to disallow the initial formation of vapor pockets. The collapse of the bubbles is nonspherical, being more correctly likened to an intense micro-jet (Karimi and Avellan 1986).

When an intense micro-jet is closely directed into the vane surface, it contributes significantly to vane surface erosion. Figure 8.3 is a photo of a centrifugal pump impeller vane inlet that has incurred substantial cavitation erosion damage. Naturally, such damage adversely affects pump energy efficiency as well as impeller structural integrity. A quite insightful visualization of the hydraulic cavitation process of a centrifugal pump vane inlet region derived from the EPFL cavitation tunnel research is shown in Figure 8.4 (Adams 2017).

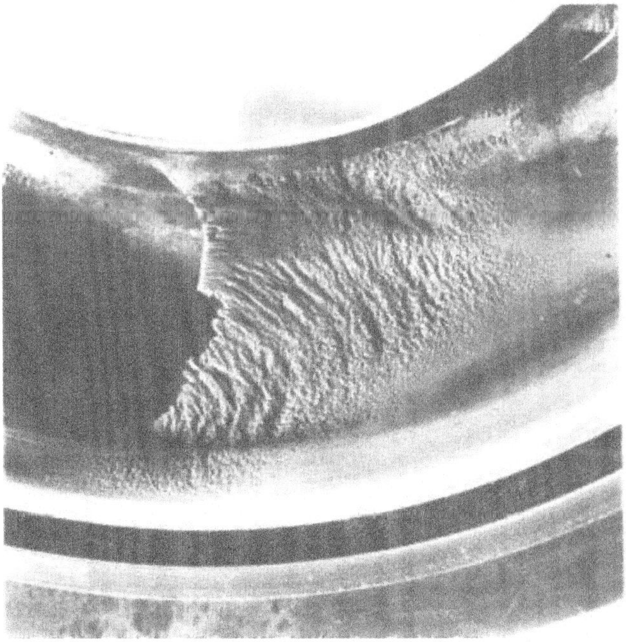

FIGURE 8.3
Cavitation erosion damage of a centrifugal pump impeller vane inlet.

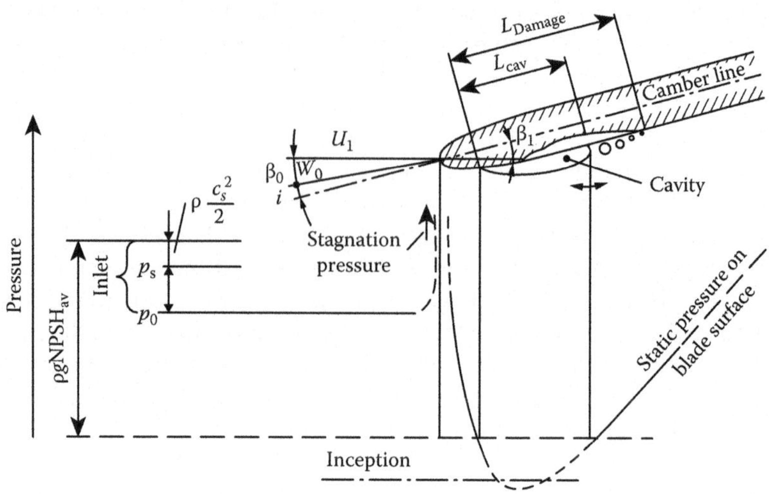

FIGURE 8.4
Pressure and erosion regions, inlet low-pressure side of impeller vane.

Cavitation-caused erosion can also occur on hydro turbine blades at the low-pressure discharge region of the turbine runner. This can occur, for example, when a turbine is operated at output power levels exceeding the hydraulic design power rating of the turbine, feasible if the unit's generator is rated above the turbine's power rating. Such an operating point above the turbine's rating might be chosen during the seasonal spring heavy-runoff river flow. The author has witnessed examples of this at hydroelectric power plants employing axial-flow Kaplan turbines where the generating income from the extra power so produced substantially exceeded the cost of periodically stainless-steel-recladding the cavitation-damaged portion of the turbine impeller blade surfaces.

Bibliography

Adams, M. L., *Power Plant Centrifugal Pumps: Problem Analysis and Trouble-Shooting*, Taylor & Francis, CRC Press, Boca Raton, FL, 182pp., 2017.

Avellan, F. and Henry, P., Theoretical and experimental study of the inlet and outlet cavitation in a model of a Francis turbine, *Proceedings of 12th IARH Symposium on Hydraulic Machinery in the Energy Related Industries*, Stirling, Scotland, pp. 38–55, 1984.

Avellan, F. and Henry, P., Towards the prediction of cavitation Erosion: IMEHF research program, *Proceedings EPRI Symposium on Power Plant Pumps*, New Orleans, LA, pp. 1–22. 1987.

Avellan, F., Henry, P., and Ryhming, I., A new high-speed cavitation tunnel for cavitation studies in hydraulic machinery, *Proceedings, ASME Annual Meeting*, Boston, MA, Dec. 1987.

Favre, J. N. and Walther, W., Analyse de la cavitation d'entrée d'um aubage Kaplan par l'application d'une method de calcul inverse bi-dimensionnelle, *Procedings of the 13th IARH Symposium on Progress in Technology*, vol. 1, pp. 4.1–4.14, 1986.

Karimi, A. and Avellan, F., Comparison of erosion mechanisms in different types of cavitation, *Wear*, 13, 305–322, 1986.

9

Swiss Federal Institute Turbomachinery Lab at ETH Zurich

This chapter is primarily devoted to the legacy of my late friend Professor Georg Gyarmathy, turbomachinery professor at the Swiss Federal Institute ETH Zurich (1984–1998). Professor Gyarmathy was the fourth holder of the ETH turbomachinery professorship, succeeding in 1984 the then retiring Professor Walter Traupel. The first holder of this prestigious ETH chair was Aurel Stadola, author of the pioneering book *Die Dampfturbine* (*The Steam Turbine*) in 1903, with one of his ETH students being Albert Einstein.

Georg's maternal uncle was the well-known Hungarian engineer Gyorgy Jendrassik, who did pioneering work in development of diesel engines and gas turbines. It was the author's good fortune to be invited by Professor Gyarmathy in 1986 as a guest professor at the ETH to give a semester course on the *Rotor Dynamics of Turbomachinery*. Over the following years, the author gave a number of short courses and invited lectures at the ETH, where he was first inspired to write his book, *Rotating Machinery Vibration* (2000, 2010).

Georg graduated with honors from the Budapest University of Technology in 1956. Later that year, following the Hungarian revolution, he fled with a sister to Switzerland by way of Austria, traversing still live mine fields left over from a prior conflict. He worked at Brown Boveri Corporation (BBC) in Baden, Switzerland from 1956 until 1959, when he started at the ETH to work on his doctoral thesis. After earning his doctoral degree from the ETH, he returned to work at BBC for several years. At BBC, he became head of the Large Steam Turbine development testing department and, subsequently, head of the BBC Turbocharger Division. He left BBC in 1984 to become the ETH turbomachinery professor. In 1991, he received an honorary doctoral degree from the Budapest University of Technology. Upon his retirement from the ETH in 1998, he was appointed Consul General in charge of the Consulate of Hungary in Munich, Germany, a position he held until 2003. In the fall of 2009, Georg passed away suddenly while visiting Kenya.

9.1 Low-Pressure Condensing Steam Turbine Research

Professor Gyarmathy made a major impact in the field of steam turbines quite early in his career with his ETH doctoral thesis (Gyarmathy 1962). In the early twentieth century, the development of large steam turbines to drive electric power generators, all the now-recoverable energy from the steam by turbine stages was not utilized because of the inherent problem of damage to turbine blades if the steam starts to condense into sizable water droplets before exiting the turbine. Even today, the rare but serious operating control malfunction that allows water backflow into the low-pressure (LP) turbine (referred to as a *water event*) will do catastrophic impact damage to the at-speed LP steam turbine blades. Benefitting from post-WWII research into the aerodynamics and coupled heat transfer phenomena involved in the condensing steam turbine, large modern LP turbines have been engineered to function with the last LP stage operating below the condensation pressure of the steam at its temperature. How is that possible?

The author's own nonspecialist explanation of how that is possible follows. Since condensation involves heat transfer, it is not an instantaneous phenomenon. So the success of the modern so-called condensing LP steam turbine, operating with a discharge pressure less than the condensation pressure, is that the fast-moving steam passing through the LP turbine does not "have enough time" to condense fast enough to form damaging-sized droplets before it exits the LP turbine. Thus, the steam in the LP turbine discharge chamber immediately downstream of the last stage of LP blades proceeds to "catch up," commencing condensation without impinging upon the blades. Professor Gyarmathy's pioneering research on condensing steam turbines is recognized as the most definitive on the subject.

9.2 Turbomachinery Stage Research

An insightful and comprehensive fundamentals-based treatment of impeller–diffuser rotating stall is given by Gyarmathy (1996), with a computational fluid dynamics (CFD) and experimental flow visualization study (File et al. 1997). That research effort continued at the ETH Turbomachinery Laboratory, combining CFD simulations with experimental tasks (Saxer-Felici et al. 1999, Gyarmathy et al. 2001). The author's own involvement with centrifugal compressor rotating stall was for a major manufacturer of large industrial air-conditioning chiller packages. That work was to computationally simulate the nonsynchronous nonlinear rotor vibration resulting from impeller-eye rotating stall using the modeling formulation code developed by Adams (1980).

Saxer-Felici et al. point out that the flow stability of any turbo compressor system becomes compromised as the mass flow rate is reduced from its nominal value below the so-called limit of compressor dynamic stability. While the well-known compressor flow-instability phenomena, *surging* and *rotating stall*, are conditions both associated with low-flow operation, they are two distinct phenomena. They may both occur at the same time, but not necessarily. Unstable-flow phenomena, while always associated with centrifugal pumping of compressible fluids, are also of serious concern in the centrifugal pumping of liquids, particularly in high-energy density pumps (Adams 2017).

The ETH test rig in this research is shown in Figures 9.1 through 9.3. With simultaneous flow visualization and time-mean unsteady static pressure measurements, in-depth comparisons were made with the in-parallel CFD simulation work. The CFD unsteady flow simulations provided insight into the visualized/measured water model test rotating stall conditions. The visualized stall patterns combined with the measured pressures enabled CFD modeling improvements. That provided improved accuracy and confidence in CFD design simulation analyses employed by manufacturers of products with compressor cascade configurations.

Further cutting-edge ETH research on compressor unstable flow is also reported in other ETH publications by Gyarmathy et al., Hunziker et al., and Ribi et al. as listed here in the chapter's bibliography. Their groundbreaking

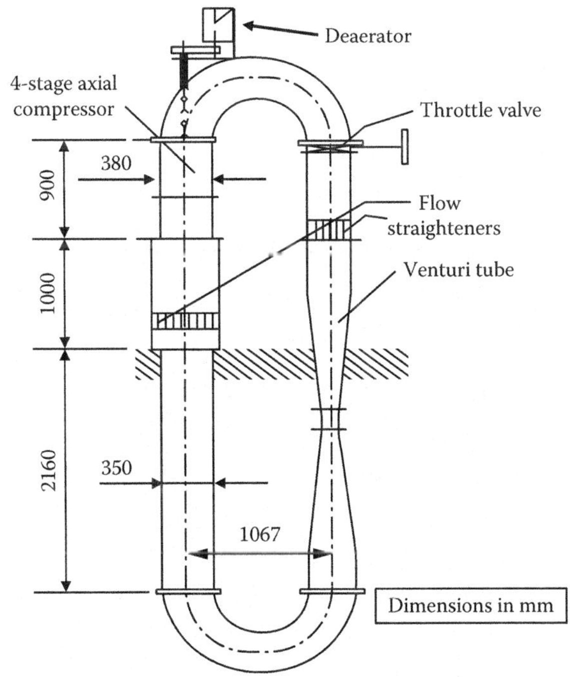

FIGURE 9.1
Schematic of water model rotating stall test rig.

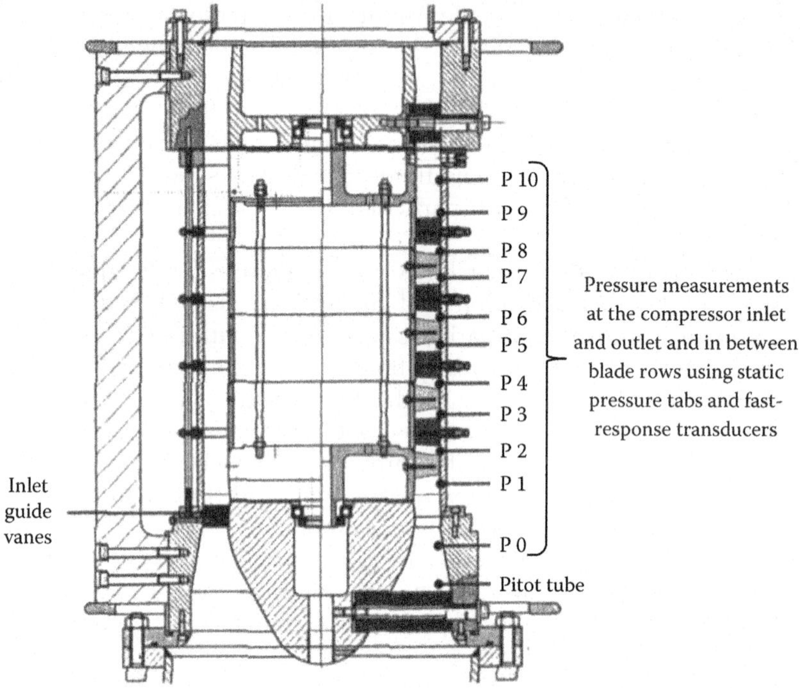

FIGURE 9.2
Pressure measurement positions in a 4-stage axial compressor model.

FIGURE 9.3
Photo of a 4-stage axial compressor water model (transparent casing).

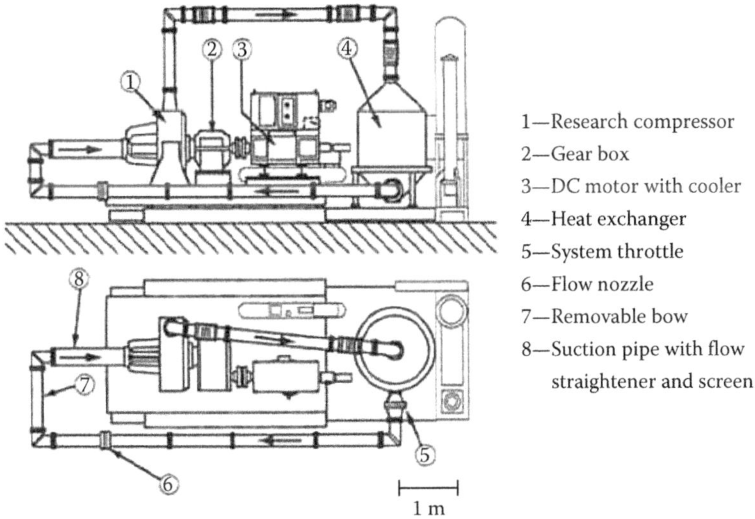

FIGURE 9.4
Surge test rig for centrifugal compressor research.

work into *mild surge* utilized the test rig illustrated in Figure 9.4. The stationary front and back walls encompassing the compressor research impeller, Figure 9.5, were instrumented with several wall pressure taps connected to high-frequency response pressure transducers, Figure 9.6. Ribi and Gyarmathy (1993) used the same test rig to investigate compressor impeller rotating stall as a trigger for mild-to-deep surge. Numerous turbomachinery

FIGURE 9.5
Centrifugal compressor research test impeller.

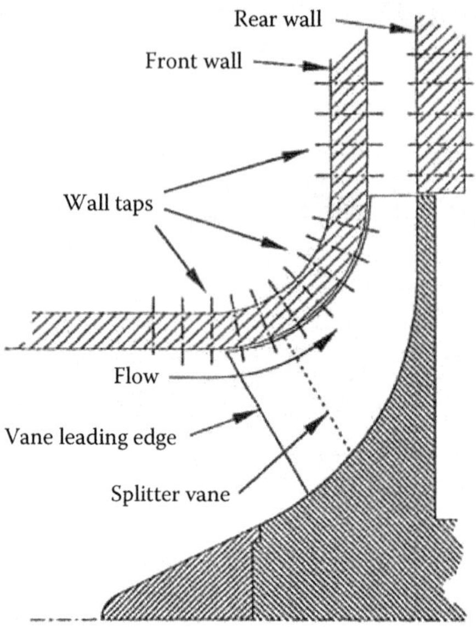

FIGURE 9.6
Location of wall pressure taps and pressure transducers.

research projects at the ETH span many past decades back to the early twentieth century during the tenure of Professor Stadola. And pioneering turbomachinery research continues at the ETH by the successors to Professor Georg Gyarmathy and his team.

Bibliography

Adams, M. L., Nonlinear dynamics of flexible multi-bearing rotors, *Journal of Sound & Vibration*, 71 (1), 129–144, 1980.

Adams, M. L., *Rotating Machinery Vibration: From Analysis to Troubleshooting*, Marcel Dekker, New York, 354pp., 2000.

Adams, M. L., *Rotating Machinery Vibration: From Analysis to Troubleshooting*, 2nd edn., Taylor & Francis, CRC Press, Boca Raton, FL, 465pp., 2010.

Adams, M. L., *Power Plant Centrifugal Pumps: Problem Analysis and Troubleshooting*, Taylor & Francis, CRC Press, Boca Raton, FL, 182pp., 2017.

File, G., Gyarmathy, G., and Staubli, T., Water model of a single stage centrifugal compressor for studying rotating stall, *Proceedings, Second EuropeanTurbomachinery Conference*, Antwerp, Belgium, 1997.

Gyarmathy, G., Grundlagen Eine Thsoeie Der Nassdampfturbine (Foundations of a theory of the wet-steam turbine), ETH (English translation, 274pp., 1964, available for down load at http://www.dtic.mil/get-tr-doc/pdf?Location=U2&doc=GetTRDoc.pdf&AD=AD0489324), 1962.

Gyarmathy, G., Impeller-diffuser momentum exchange during rotating stall, ASME Paper No. 96-WA/PID-6, New York, 1996.

Gyarmathy, G., Hunziker, R., Ribi, B., and Spirig, M., On the change of impeller flow non-uniformities with flow rate in a centrifugal compressor, *Proceedings of the Institution of Mechanical Engineer*, IMechE C423/054, Mar. 1991.

Gyarmathy, G., Inderbitzin, A., and Staubli, T., Rotating stall in centrifugal compressors, Video, Turbomachinery Laboratory of the Institute of Energy Technology, Swiss Federal Institute of Technology, Zurich, Switzerland, 1997.

Gyarmathy, G., Inderbitzin, A., and Staubli, T., Visualization of rotating stall in a full size water model of a single-stage centrifugal compressor, *La Houille Blanche*, 11(3), 40–45, June 2001.

Hunziker, R., Ribi, B., Spirig, M., and Gyarmathy, G., On the influence of different impellers on a radial compressor stage with vaneless diffuser, *Interfluid First International Congress on Fluid Handling Systems*, Essen, Germany, pp. 231–240, Sept. 1990.

Ribi, B. and Gyarmathy, G., Impeller rotating stall as a trigger for the transition from mild to deep surge in a subsonic centrifugal compressor, ASME-Paper No. 93-GT-234, New York, May 1993.

Ribi, B., Strömungsphänomene und deren Interaktion in einem Radialverdichtersystem, VDI-GET-Fachtagung, VDI Berichte Nr. 1109, Mar. 1994.

Ribi, B. and Gyarmathy, G., The behaviour of a centrifugal compressor stage during mild surge, *First European Conference on, Turbomachinery—Fluid Dynamic and Thermodynamic Aspects*, Bavaria, Germany, pp. 341–356, VDI Berichte Nr. 1186, Mar. 1995.

Ribi, B. and Gyarmathy, G., Energy input of a centrifugal stage into the attached piping system during mild surge, *ASME Journal of Engineering for Gas Turbines and Power*, 121, 325–334, Apr. 1999.

Saxer-Felici, H. M., Saxer, A. P., Inderbizin, A., and Gyarmathy, G., Prediction and measurement of rotating stall cells in an axial compressor, *ASME Journal of Turbomachinery*, 121, 365–375, 1999.

10

Axial Location and Size of Progressing Shaft Cracks

This chapter summarizes research in the author's rotating machinery laboratory at Case Western Reserve University (CWRU) for the detection in real time of developing cracks in shafts of operating machines. The objective was to extend existing shaft crack detection methods to locate the axial position of the propagating shaft crack and the size to which the crack has grown. The work was funded through a research PhD fellowship sponsored by First Energy Corporation, Akron, Ohio. This research effort was to explore new and improved methods for detecting cracks in large power plant rotating machinery, like main steam turbine generator sets, feedwater pumps, combustion gas turbines, fans, reactor coolant pumps, high-pressure safety injection pumps, and other major machinery types in both fossil and nuclear power plants. A new shaft crack research test rig design was conceived as a concluding task of this research.

10.1 State of the Art in Real Time Shaft Crack Detection

The initiation and subsequent propagation of a crack through the shaft is, of course, among the most dreaded failure types in rotating machinery. As a consequence, the early diagnosis and careful trending of vibration symptoms for cracked shafts is a well-studied topic within the machinery vibration monitoring field. Muszynska (1995) provides an insightful explanation of cracked-rotor vibration symptoms, employing a simple 2-degree-of-freedom single-mass rotor dynamics model. Muszynska's model embodies the two prominent symptoms that a rotor crack superimposes upon a simple unbalance-only vibration model. These two effects are (1) a bending stiffness reduction aligned with the crack direction and (2) a crack-local shift in the bending neutral axis (the rotor therefore bows) corresponding to the crack direction.

The first of these two effects produces a twice-rotational-speed (2*N*) vibration component. This is similar to what would occur prominently in long 2-pole power plant generators if not for the standard radial slots that are cut in such generator rotors to equilibrate the two principle bending moments of inertia (Adams 2010). The second of these effects produces a synchronous (*N*) vibration component, which vectorially adds to the pre-existing residual unbalance-driven synchronous radial vibration and changes as the crack grows.

The rotor radial vibration symptoms for a developing rotor crack are therefore (1) the emergence and growth of a 2*N* vibration component simultaneous with (2) the emergence of a progressively changing synchronous vibration amplitude and phase angle from the rotor-bow-induced unbalance. An additional symptom is apparent when rotor *X* and *Y* vibration orbital displacements are monitored. That is, the rotor vibration orbit will appear similar to the typical orbit with *N* and *N*/2 harmonics superimposed (Adams 2010), except that the period of the cracked shaft orbit is one revolution (not two) as immediately detectable from the presence of only one keyphasor mark per period of orbital vibration. There have been some remarkably accurate predictions of how long a rotor can operate before it fails from a sudden through fracture precipitated by shaft crack propagation. In one well-substantiated case, the supplier of the vibration monitoring system (Dr. Donald Bently of Bently-Nevada Co.) predicted the exact number of operating days remaining before the failure of a PWR nuclear reactor *primary coolant pump shaft* (Byron-Jackson) in a PWR commercial electric power-generating plant (Crystal River, Unit-4, Florida Power Corp.). That nuclear plant's operators were unfortunately skeptical of Don Bently's prediction. After the reactor coolant pump shaft failed as predicted to the day by my friend Don, they became converted "believers," albeit too late. Due to subsequent, much more serious failure and major containment vessel damage, that nuclear power unit is no longer in service.

10.2 Basic Principles of the New CWRU Approach

The new approach in its initial development phase was to detect both the depth and axial location of radial propagating cracks in rotating shafts. Use of accelerometers was identified as the ideal detection and measurement sensor, based on the use of standard accelerometers to capture the shaft-end compression waves from a crack's once-per-revolution closure. An application of the classic one-dimensional Wave Equation to estimate the shaft compression waves was combined with the mass of the shaft-end accelerometer to provide an estimate of the measurable axial shaft-end acceleration pulses caused by a specified shaft crack closing. A general equation to estimate the associated shaft end acceleration pulse magnitudes starts with the classical 1D Wave equation as follows:

$$\frac{1}{c^2}\frac{\partial^2 u}{\partial t^2} = \frac{\partial^2 u}{\partial x^2} \tag{10.1}$$

where

$c = \sqrt{E/\rho}$ is the wave propagation speed
E is the modulus of elasticity
ρ is the mass per unit volume
$u(x, t)$ is the axial wave displacement
x is the axial position

This wave-theory-based starting point postulates that as a shaft crack opens and closes suddenly each revolution, the impact-like closing of the shaft crack produces an axially traveling compression wave in both axial directions from the crack. This, of course, requires that there is an axially distributed bending moment present in the shaft. This a reasonable supposition since any bearing, gear, hydraulic, or other radial support forces holding the shaft in its running static equilibrium position will produce a nonrotating bending moment distribution.

Since the shaft rotates with this imposed nonrotating beam-bending stress distribution, a turning-speed stress reversal is imposed upon the rotating shaft. Thus, there is always the danger of a fatigue microcrack forming along the direction of the maximum alternating shear stress, which is approximately 45° angled with respect to the axial beam-bending normal stress (ala Mohr's Circle), which is maximum at the shaft's outer diameter. The microcrack then redirects at approximately 45°, propagating normal to the maximum alternating beam-bending normal stress, $\sigma = Mc/I$. So, the crack opens and closes at each revolution of the shaft. Since the sharp tip of the propagating crack is a concentrated stress raiser, the crack grows a little at each revolution when the crack opens. This is prevented only if there is a sufficiently large superimposed axial static compressive force (normal stress) on the shaft to prevent the crack from opening, a nontypical condition. The shaft fails when the propagating crack destroys enough of the shaft, reducing the cross-sectional load-carrying area to below that needed to support the nonrotating bending stresses. The failure then climaxes in a sudden complete through fracture of the shaft. This topic is covered in detail in any major undergraduate machine design course.

10.3 Research Test Rig

Although the grant from First Energy Corporation did not include funding to build a new and unique test rig, an appropriate shaft crack research test rig was conceived of as a concluding task of this research grant. That rig design is illustrated in Figure 10.1.

Following the completion of the test rig design configuration, the focus next was on identifying the type of sensor at the axial ends of the shaft that would be best suited to capture the traveling elastic wave signals emanating from the shaft crack when it slams shut once at each revolution. This led to a major breakthrough in the research.

After researching various potential sensor options on feasibility, Kistler piezoelectric accelerometers were selected because they are quite accurate and the CWRU lab already had a wide g-level range of these. However, the axially traveling elastic velocity wave as predicted from first principles, Equation 10.1, yields the classic reverse-direction wave reflection as the wave reaches the end of the shaft, which theoretically means infinite acceleration, if only for an infinitesimal axial sliver of mass. So, the prior analytical treatment (LaBerge and Adams 2007) did not provide a direct means for estimating the measurable axial shaft acceleration spike as the wave reaches the ends of the shaft.

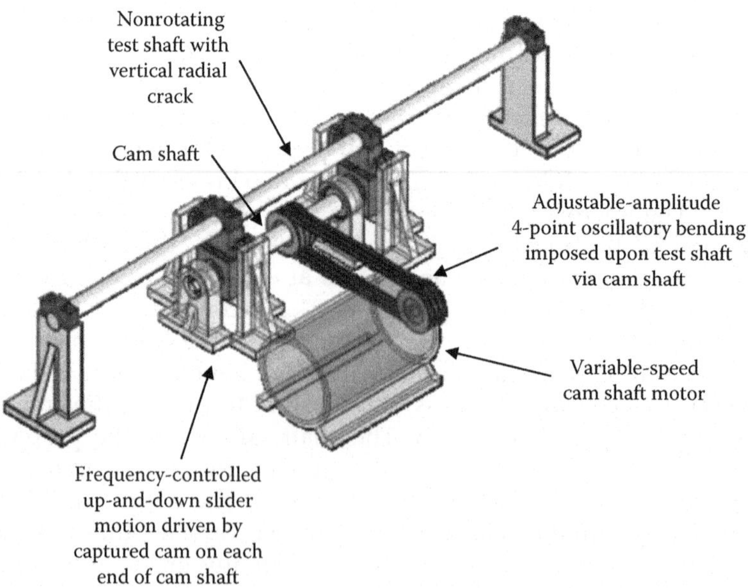

FIGURE 10.1
Shaft crack research test rig configuration.

Going "back to the drawing board," we devised an approach to estimate the measurable axial shaft acceleration spike as the wave reaches the end of the shaft. The approach is as follows: If the free end of the shaft bar section were instead constrained, the peak force exerted on the axial constraint would be as follows:

$$F = \sigma A = \varepsilon EA = EA\frac{\partial u}{\partial x} = \left(\frac{\pi D^2 E}{4}\right)\frac{\partial u}{\partial x} \tag{10.2}$$

where

σ is the traveling-wave shaft axial stress
ε is the corresponding axial strain

This value of F is then used with $F = ma$ to estimate acceleration of the accelerometer and its magnetic attachment base. For example, a 25 g accelerometer with its magnetic base typically weighs about 28 gm. The bar compression displacement should be equal to the bar velocity V multiplied by the time t_p it takes for the wave to propagate from the crack to the bar's free end, that is, the displacement of the crack in the bar relative to the free end over the time period t_p. The corresponding strain is then given by the following equation:

$$\varepsilon = \frac{Vt_p}{L} \tag{10.3}$$

where L is the axial length from the crack to shaft's free end. So the shaft end acceleration is estimated by the following equation:

$$a = \frac{F}{m} == \frac{\pi D^2 EVt_p}{4mL} = \frac{\pi D^2 EV}{4mc} = \frac{\pi D^2 V\sqrt{\rho E}}{4m} \tag{10.4}$$

where

D is the shaft OD
$c = \sqrt{E/\rho}$ is the wave propagation speed
m is the mass of the accelerometer with its magnetic base plus any additional mass added, if needed, to attenuate the measured acceleration spike

As explained by LaBerge and Adams (2007), the bar velocity V is a function of shaft diameter, crack depth, and shaft rotational speed. Thus, an appropriate research test rig is needed to calibrate shaft-end acceleration and velocity as a function of shaft diameter and crack depth. Subsequent FEA modeling can then extrapolate test results to other shaft sizes and configurations.

Running numbers through the derived equations for a wide range of shaft diameters and crack depths resulted in a "eureka" moment. Our numbers showed that one can adjust the shaft-end-attached mass to achieve measurable optimum acceleration spikes for virtually any expected shaft diameter and crack depth, including early crack depths. The first-principle-based estimates for a measurable shaft-end acceleration spike at last provided a major breakthrough. In summary, the author can state with confidence that this new shaft crack detection method is a quite promising approach to provide early shaft crack detection along with crack axial location and crack depth based on comparative wave arrival times and intensities.

Bibliography

Adams, M. L., *Rotating Machinery Vibration: From Analysis to Troubleshooting*, 2nd edn., Taylor & Francis, CRC Press, Boca Raton, FL, 465pp., 2010.

LaBerge, K., Shaft crack detection from axially propagating stress waves of crack closings under rotation, PhD dissertation, Case Western Reserve University, Cleveland, OH, 2009.

LaBerge, K. and Adams, M. L., Analysis of the elastic wave behavior in a cracked shaft, *Proceedings, ASME 2007 International Design Engineering Technical Conferences*, DETC2007-35305, LasVegas, Nevada, Sept. 2007.

Muszynska, A., Vibrational diagnostics of rotating machinery malfunctions, *International Journal of Rotating Machinery*, 1 (3–4), 237–266, 1995.

11

Cleveland State University Wind Turbine Tower

The pioneering development summarized in this chapter was undertaken by Dr. Majid Rashidi, Professor of Mechanical Engineering at Cleveland State University (CSU), Ohio. The fundamental underpinning of this development is the well-known amplification of free stream flow velocity near a circular cylinder within the flow field. Figure 11.1 graphically illustrates this velocity amplification effect (Rashidi et al. 2015). This work was funded by the DOE program for Energy Efficiency Renewable Energy (EERE) Wind and Water Power.

11.1 Amplified Velocity Significantly Increases Wind Power

To utilize the velocity amplification phenomenon to drive wind turbines, the Swift™ configuration shown in Figure 11.2 was chosen. Being optimized for small rooftop wind turbines, it was found to be ideally sized for operation within the velocity amplification zone of available-sized-diameter cylinders such as water towers.

CFD analyses of ideal flow around a cylinder yield the near-cylinder flow velocity to be about 1.5 times the upstream mean flow velocity, that is, a *velocity amplification factor* of 1.5. The power output, P, of a single wind turbine can be estimated by Equation 11.1.

$$P = \frac{1}{2}\rho A C_{OP} v^3 \eta_g \eta_b, \tag{11.1}$$

where

ρ is the air density
A is the turbine area exposed
C_{OP} is the coefficient of performance
v is the wind velocity
η_g is the generator efficiency
η_b is the gearbox/bearing efficiency

Clearly from Equation 11.1, a velocity amplification factor of 1.5 produces a *power amplification factor* of 3.4, the cube of the wind amplification factor.

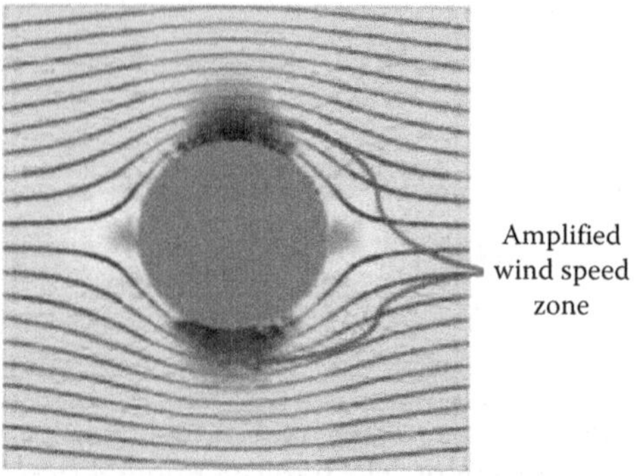

FIGURE 11.1
Ideal flow boundary streamlines around a circular cylinder.

FIGURE 11.2
Swift™ rooftop wind turbine, 82 inches diameter, 1.65 kW.

11.2 Full-Scale Prototype Installations

The first full prototype application was installed on a CSU campus building rooftop, Figure 11.3, employing four of the Swift wind turbines shown in Figure 11.2. It met all performance expectations. The demonstrated success of this first full-size prototype provided significant justification for future

FIGURE 11.3
CSU campus full-scale wind turbine tower.

installations on the numerous water towers and other similar structures at power plants, oil refineries, etc., typified by those shown in Figure 11.4.

The next and more advanced second-generation wind turbine tower configuration developed by the CSU team was installed with considerable national publicity at the Cleveland Indians baseball stadium. It utilized a

FIGURE 11.4
Structures for potential installations of multiple small wind turbines.

FIGURE 11.5
Scale model of a second-generation wind turbine tower.

(a) (b)

FIGURE 11.6
Cleveland Indians stadium installation: (a) illustration and (b) photo.

more sophisticated wind collection geometry, Figure 11.5, that increased the wind velocity amplification factor even higher than that of the prior cylindrical tower pictured in Figure 11.3. It employed four 2-meter-diameter turbines as illustrated and pictured in Figure 11.6. All the configurations presented here are detailed in the four U.S. patents listed in the bibliography.

Bibliography

Rashidi, M., Wind spires as an alternative energy source, Final Technical Report, U.S. Department of Energy (DOE) Award No.: DE-FG36-08GO88016, October 2012.

Rashidi, M., Kadambi, J. R., and Ebiana, A., Performance of a rooftop wind turbine system having a wind deflecting structure, experimental results, *Proceedings, IMECE2015, ASME International Mechanical Engineering Congress & Exposition*, Houston, TX, November 2015.

Rashidi, M., U.S. Patents Nos. 7,540,706 B2 (June 2, 2009), 7,679,209 B2 (March 16, 2010), 7,679,209 B2 (March 16, 2010), 8,002,516 B2 (August 23, 2011).

12

Compressor Refrigerant–Oil Separation Seal

A new impeller-to-sump seal was developed for a line of hermetically sealed high-speed (34,000 rpm) refrigerant centrifugal compressor units. By using an actual compressor unit as the heart of the development test rig, experiments were designed to simulate actual in-service operating conditions. Both, the new seal configurations and the original seal were tested. A new seal configuration that resulted in a 25:1 reduction in refrigerant leakage met all other technical and economical requirements dictated by the application. This seal development work is documented in full detail by Adams and Raimondi (1977).

The specific line of liquid chiller units here ranged in capacity from 80 to 240 tons. Units from 80 to 116 tons employed one centrifugal compressor unit. On larger capacities up to 240 tons, two compressors were employed. The complete hermetically sealed unit operating with R-12 included a condenser and a chiller. Figure 12.1 shows both the single-compressor and double-compressor configurations.

12.1 Liquid Chiller Centrifugal Compressor, Seal, and Motivation

A cutaway of the centrifugal 34,000 rpm compressor is illustrated in Figure 12.2. It is driven by a 3550 rpm electric motor through a single-pinion speed increaser gear set, nominally at 10:1 speed ratio. The compressed refrigerant gas is piped to the condenser (heat exchanger), which removes heat from the compressed gas, condensing the refrigerant and rejecting heat to the cooling tower water. The liquid refrigerant then passes through a rapid expansion device, which reduces its temperature and pressure, returning to gaseous form, which is piped to the chiller. The chiller (heat exchanger) adds heat to the low-pressure, low-temperature refrigerant gas, removing heat from the house-cooling water. The refrigerant gas is then returned to the suction side of the compressor (impeller eye), completing the cycle, Figure 12.3.

The single-stage compressor needs two high-speed dynamic seals. One seal separates the compressor discharge from its inlet suction side and is a standard stationary labyrinth bushing, which closely encircles the outside diameter of the impeller eye. The other dynamic seal separates the high-pressure discharge

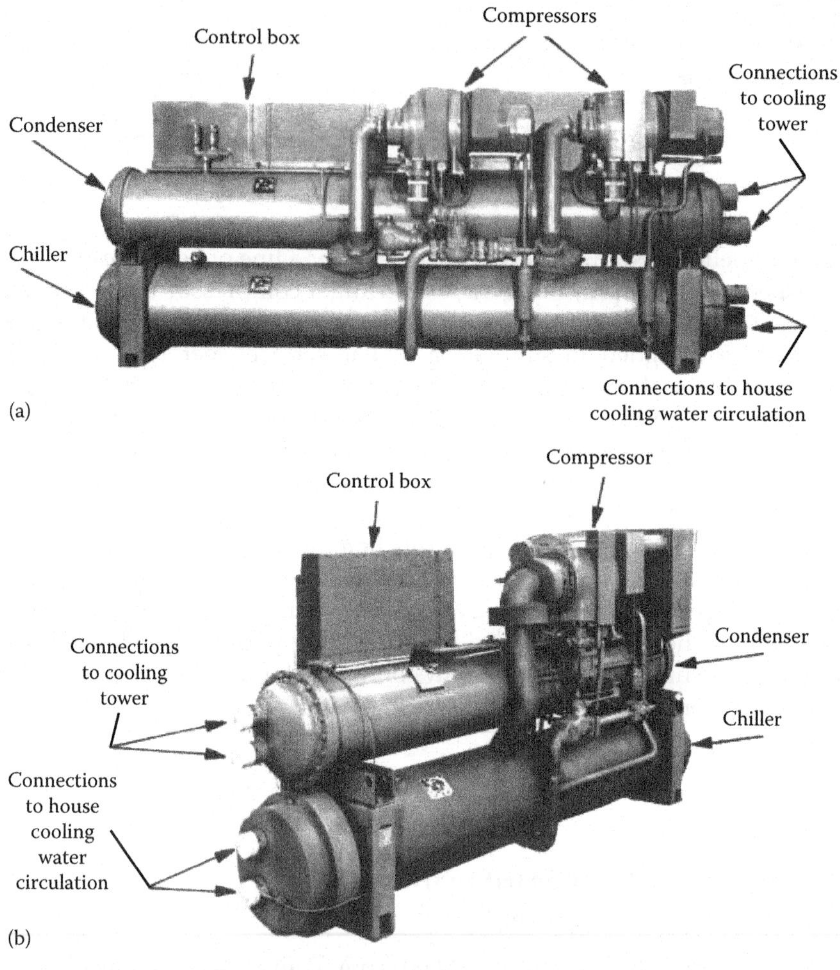

FIGURE 12.1
Water chiller, (a) double-compressor unit, (b) single-compressor unit.

chamber of the compressor from the speed increaser gear chamber/sump, Figure 12.3, the one for which this product improvement effort was undertaken. New seal ideas were generated (Adams 1975, Adams and Raimondi 1979) that justified investing in a serious experimental effort. The major challenges were (1) to devise a test rig that would accurately reproduce the seal's actual compressor environment and (2) accurately measure the seal leakage from the compressor discharge chamber to the speed increaser gear chamber/sump.

As previously stated, the entire chiller package is hermetically sealed. So, the overriding motivation was to drastically reduce the high-pressure seal leakage to the speed increaser gear chamber/sump. That is because whatever oil gets transported over into the refrigerant loop must be periodically

FIGURE 12.2
Refrigerant high-speed centrifugal compressor, 34,000 rpm.

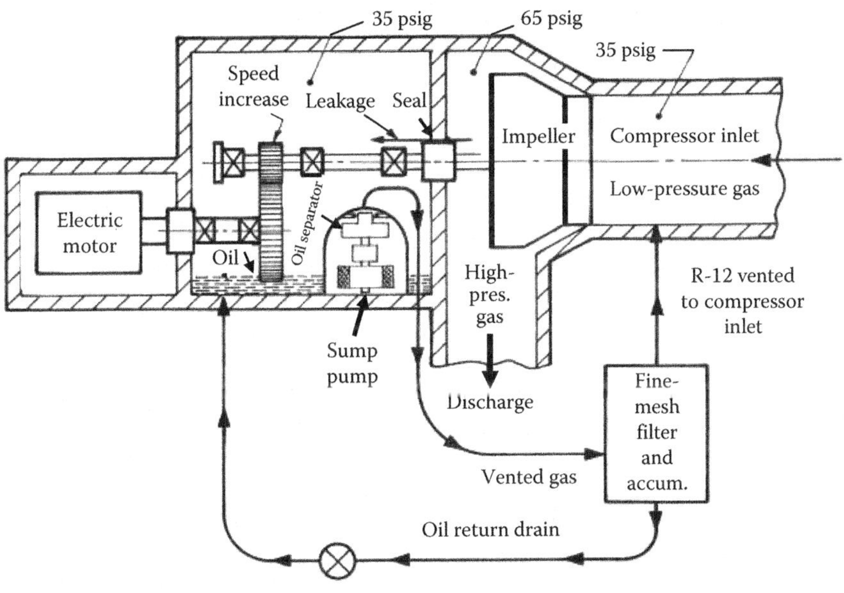

FIGURE 12.3
Schematic of a compressor unit.

drained back into the gear chamber sump. Otherwise, the heat exchanger tubes in the refrigerant loop would acquire an oil coating, producing a significant reduction in the heat exchanger efficiency. This necessitated the chiller unit to be shut down for about 15 minutes about every two days during full-capacity cooling periods. That required a maintenance person to manually shut down and subsequently restart the unit—a product shortcoming.

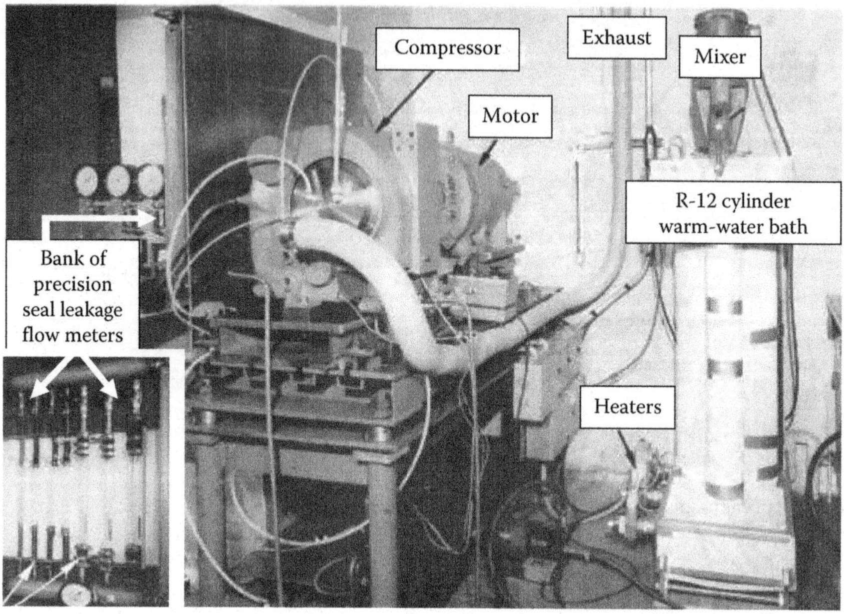

FIGURE 12.4
Test rig.

A picture of the completed test rig and a schematic illustration of it are shown in Figures 12.4 and 12.5, respectively. To achieve compressor speeds up to the 34,000 rpm operating speed, an actual centrifugal compressor was scavenged for the test rig. Of course, the impeller vanes were machined off as illustrated in Figure 12.5. As Figure 12.5 shows in detail, R-12 heating and pressure regulation means were employed to create the temperatures and pressures delivered from the bottled R-12 to replicate operating conditions of the seal in the actual application. Superimposed on the Figure 12.4 photo is the frontal view of the bank of several high-precision rotameter flow meters used to measure seal leakage. Several flow meters covering a wide range of maximum flow were employed so that one could be selected to yield a near-full-scale—thus most accurate—reading no matter how small the seal leakage rate.

12.2 Seal Configurations Tested

The seal configurations tested evolved as the project progressed, starting with the existing seal configuration already employed in the product, Figure 12.6. The first configuration tested was the product's original seal, shown in Figure 12.6. The original seal was a simple cylindrical rotating

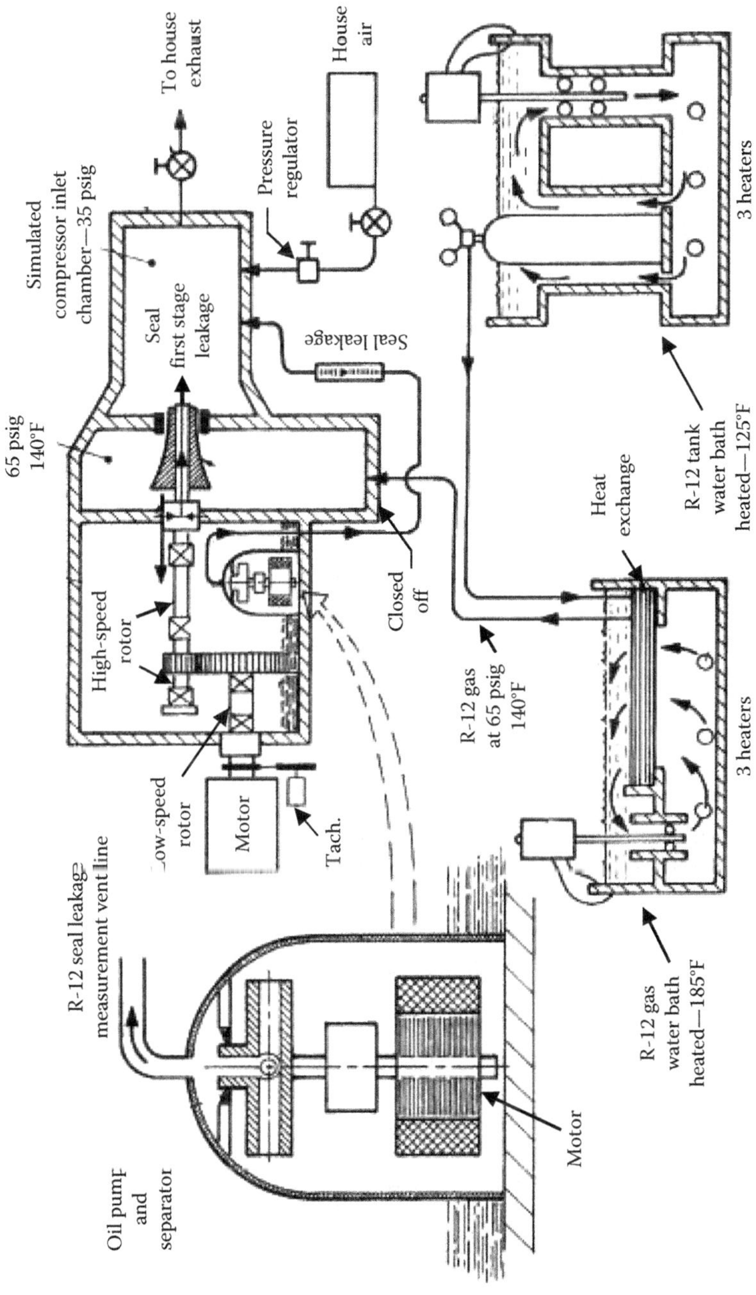

FIGURE 12.5
Schematic of a test rig.

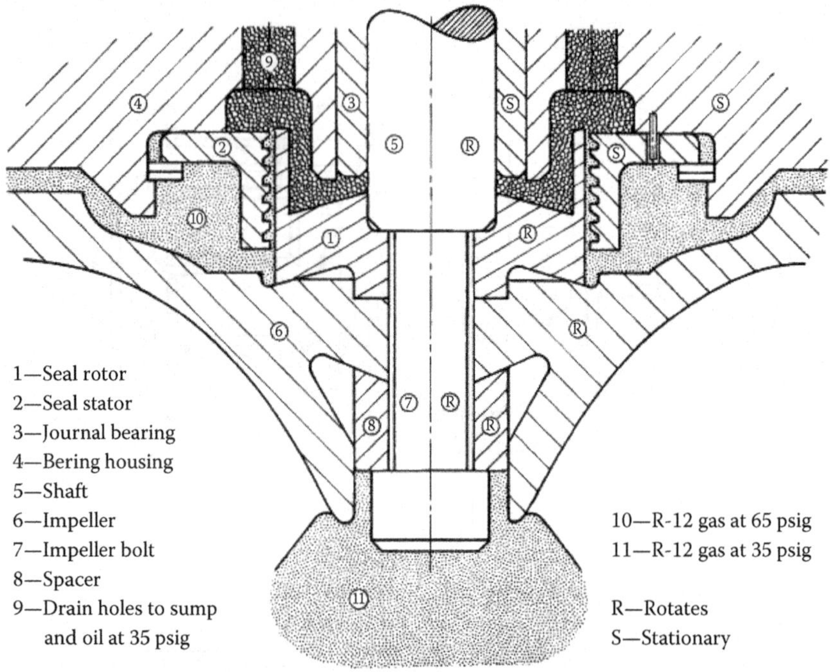

FIGURE 12.6
Original seal.

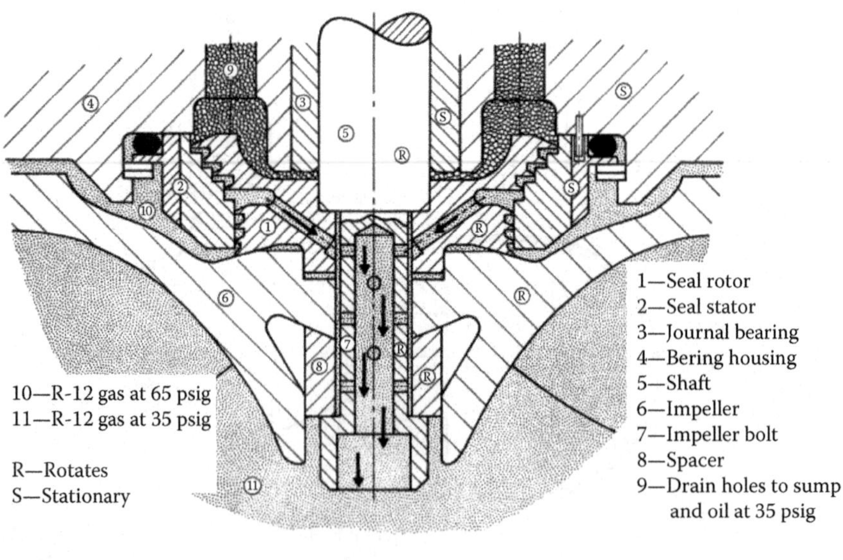

FIGURE 12.7
Oblique-labyrinth, nonscavenging vent seal. (From Adams, M.L., Rotating element fluid seal for centrifugal compressor (Two (2) U.S. patents of the same title), U.S. Patent Nos. 3, 927, 889 and 3, 927, 890, 1975.)

piece closely circumscribed by a nonrotating non-sharp-tipped-tooth labyrinth bushing.

The author's first concept tested for a reduced-leakage seal is shown in Figure 12.7. It was a two stage sharp tipped labyrinth (rotating) piece circumscribed by a carbon graphite size-to-size fitted nonrotating piece. The annular chamber between the two seal stages vents R-12 back to the impeller eye, which was expected to yield a near-zero pressure difference across the seal's second stage. That was expected to yield a near-zero R-12 leakage into the gear chamber, with the oblique feature of the second stage preventing any higher-density oil from "sneaking" directly into the refrigerant side of the seal. However, leakage flow test results showed an unfavorable rotor speed effect, Figure 12.8. This was evidence that the rotating seal rotor radial vent holes were producing their own "little" centrifugal compressor effect.

So, the next tested version of the vented seal concept replaced the seal rotor radial vent holes with a midstage annular chamber as illustrated in Figure 12.9. That eliminated the unfavorable speed effect of the Figure 12.7 configuration shown in Figure 12.8. The seal midstage annular chamber was vented through a few axial holes in the impeller as shown, returning the impeller retaining bolt to full integrity as well. The benefit of this configuration had an added favorable centrifugal effect of the exiting vent holes in the impeller shroud as Figure 12.9 shows. That feature countered the slight

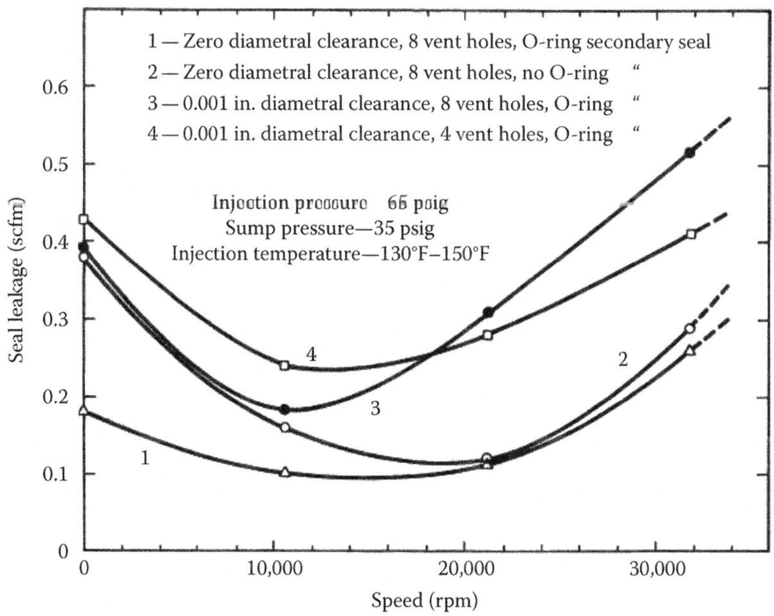

FIGURE 12.8
Leakage measured on an oblique-labyrinth, nonscavenging vent seal.

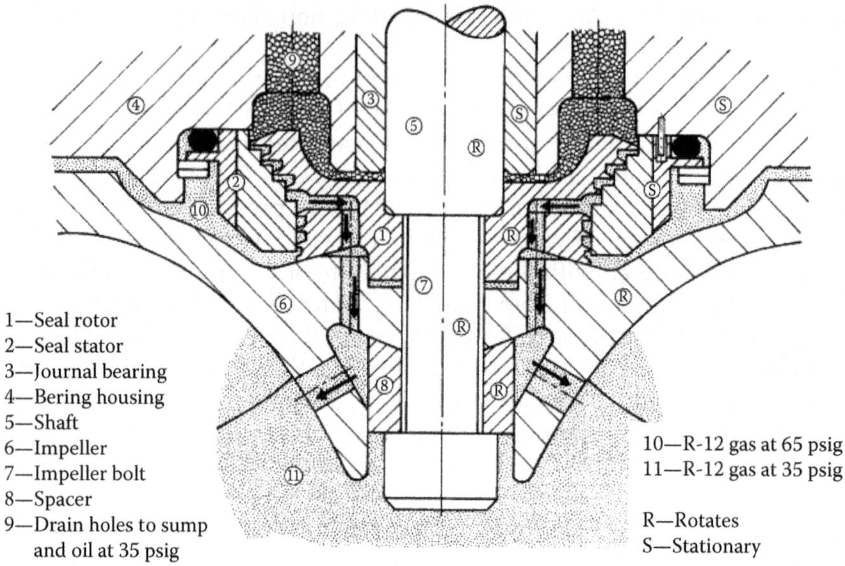

FIGURE 12.9
Oblique-labyrinth, scavenging vent seal. (From Adams, M.L. and Raimondi, A.A., Rotating element fluid seal for centrifugal compressor, U.S. Patent No. 4, 132, 416, 1979.)

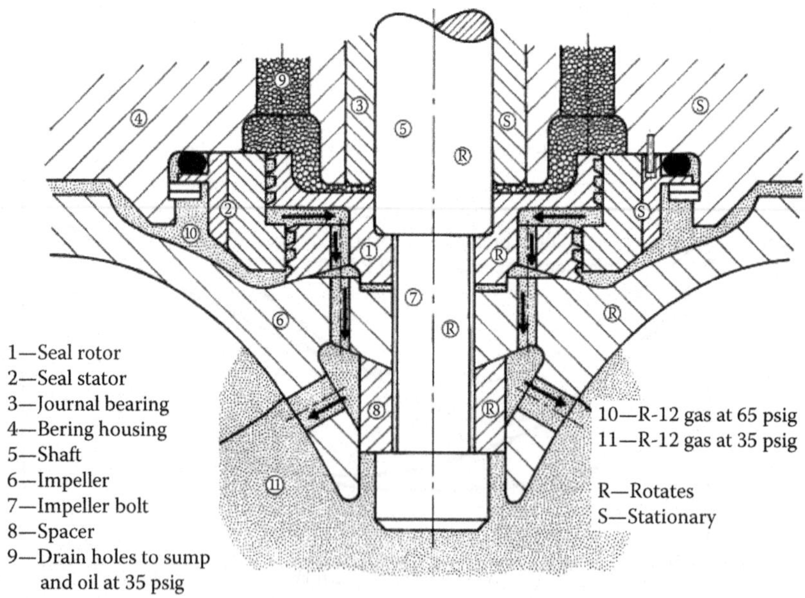

FIGURE 12.10
Parallel-labyrinth, scavenging vent seal. (From Adams, M.L. and Raimondi, A.A., Rotating element fluid seal for centrifugal compressor, U.S Patent No. 4, 132, 416, 1979.)

adverse centrifugal effect probably produced by the rotating vaneless annulus in the Figure 12.9 seal rotor. Seal leakage test results presented in the next section (Adams and Raimondi 1977) clearly demonstrate how this evolution in seal development was productively guided by the development testing. The seal configuration illustrated in Figure 12.10 was simply less expensive to produce than the Figure 12.9 version. It has the second stage diameter larger than the first stage diameter to also prevent any higher-density oil from "sneaking" directly from the gear chamber into the refrigerant side of the seal.

12.3 Test Results Provide Product Improvement

Tests on the original seal exposed its excessively high leakage rate, Figure 12.11. Those measurements showed the expected seal leakage as a function of pressure drop across the seal. This was as expected for an annular orifice flow path. The original seal's high leakage rate clearly showed why operating units in the field needed to be regularly shut down for 15 minutes every couple of days to drain oil from the refrigerant side back to the gear chamber side of the unit. Those tests also showed that there was no influence of rotor speed on the original seal's leakage characteristics.

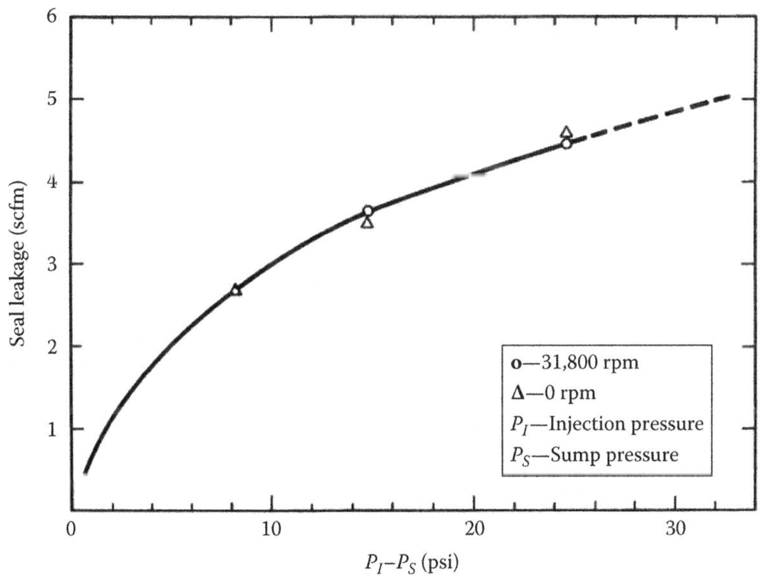

FIGURE 12.11
Measured leakage of the original seal vs pressure drop and speed.

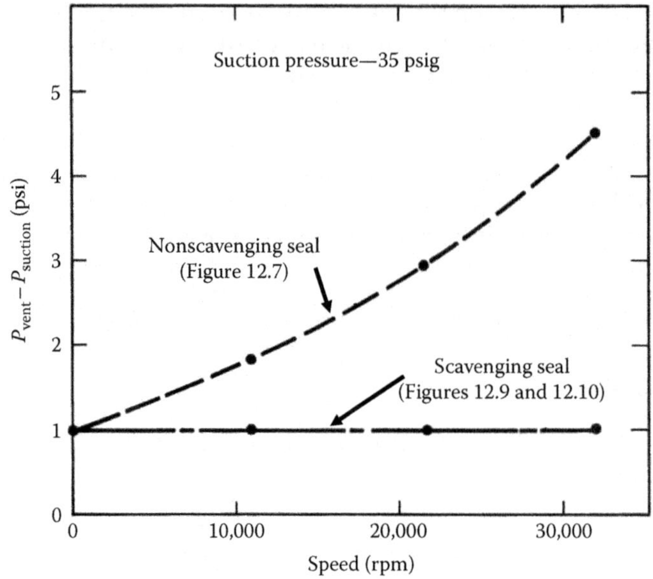

FIGURE 12.12
Measured vent pressure vs speed of the Figure 12.7 configuration.

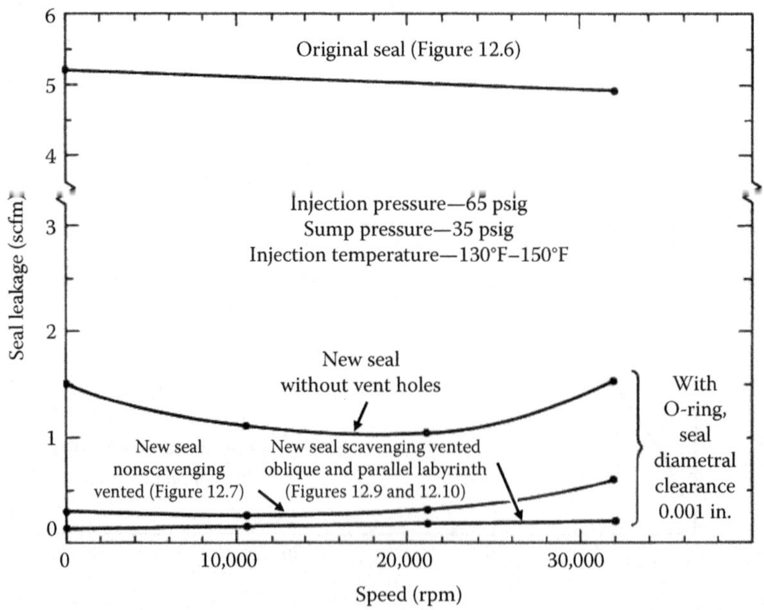

FIGURE 12.13
Measured leakage vs speed of all seal configurations tested.

Figure 12.12 shows measured leakage vs speed characteristics in the Figure 12.7 configuration. A definite speed effect was discovered that reduced seal leakage up to about the 12,000–14,000 rpm range and then increased leakage significantly up to operating speed. Pressure taps were then installed to measure the seal vent pressure, confirming the adverse centrifugal pumping effect of the seal vent holes, Figure 12.12. This measurement gave birth to the scavenging vent configurations shown in Figures 12.9 and 12.10. Seal leakage vs speed characteristics of all seal configurations tested are plotted in Figure 12.13. The comparisons are dramatic. The vented scavenged configurations yielded the lowest net leakage, and the lower-cost parallel labyrinth Figure 12.10 version was even slightly better than the oblique-labyrinth configuration of Figure 12.9. The net result deliverable of this development testing was a 25:1 reduction in seal leakage with an incremental product cost increase of only $20, that is, the cost comparison between the original seal ($5) and the Figure 12.10 configuration ($25) on a product with a sales price of several thousand dollars.

Bibliography

Adams, M. L., Rotating element fluid seal for centrifugal compressor (Two (2) U.S. patents of the same title), U.S. Patent Nos. 3, 927, 889 and 3, 927, 890, 1975.

Adams, M. L. and Raimondi, A. A., A centrifugal compressor seal, *Transactions American Society of Lubrication Engineers (ASLE)*, 20 (4), 287–294, 1977.

Adams, M. L. and Raimondi, A. A., Rotating element fluid seal for centrifugal compressor, U.S Patent No. 4, 132, 416, 1979.

13

Combined-Impeller Turbine-Driven Pump

A novel mid-1960s product development project provided the author an early career experience opportunity in performance-testing hydraulic turbomachines. The new product consisted of a centrifugal pump impeller driven by a radial outflow reaction turbine stage integrally attached to the pump impeller back shroud. The new product was developed as a more compact and higher-efficiency alternative to the conventional jet pump, Figure 13.1. In this particular project, the prototype tested was considered to be the smallest practical size of the unique combined-impeller turbine-driven pump configuration.

As illustrated, a jet pump does not need moving parts and is essentially just a *configuration of piping*. Stepanoff (1957) also shows a jet pump version employing a controlled adjustable nozzle, that is, one simple position-adjustable part to set the orifice flow area from full open to closed, similar to a Pelton impulse hydro turbine adjustable nozzle. Jet pumps are employed in various low-pressure pumping applications where, for example, high-pressure motor-driven pumps are nearby on-site, like naval ship fire pumps. As can be deduced from the Figure 13.1 example, a nearby high-pressure source, for example, a pump, provides high-pressure flow to the jet pump's high-discharge velocity nozzle, producing a jet flow that imparts its momentum to the liquid in the low-pressure tube of the jet pump. A jet pump occupies a significant longitudinal space. That size issue was the initial motivation for the development of the then-new combined-impeller turbine-driven pump product here described. Stepanoff provides a thorough treatment on the performance of jet pumps and points out that jet pumps are subject to cavitation just like centrifugal pumps when the suction head is insufficient to prevent the formation of vapor pockets in the inlet suction flow. With a fixed nozzle, efficiency vs flow tops out at less than 25%, and with a variable nozzle—35%. In comparison, the tested turbine-driven pump presented here has a peak efficiency of 45% and was significantly smaller than its jet pump counterpart.

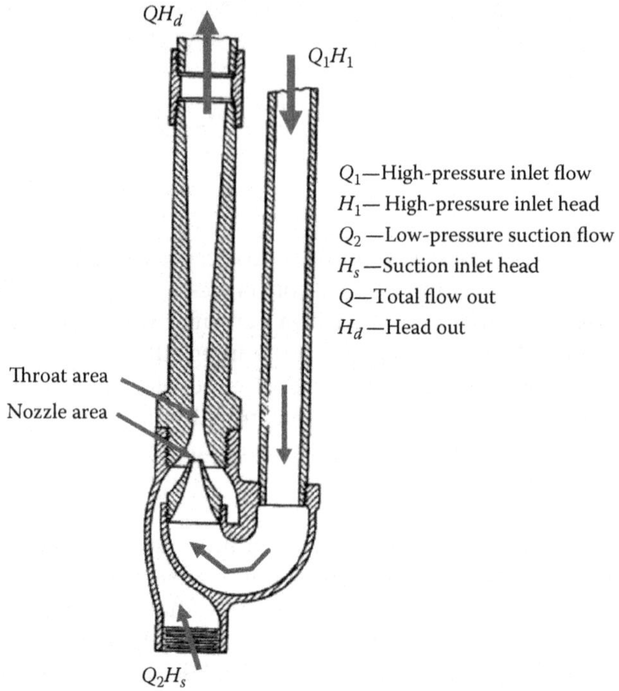

FIGURE 13.1
Jet pump with nonadjustable nozzle. (From Stepanoff, A.J., *Centrifugal and Axial Flow Pumps*, 2nd ed., Wiley, 462pp., 1957.)

13.1 Description of the Combined-Impeller Turbine-Driven Pump

The development unit tested is shown in Figure 13.2. Its compactness is obvious. As shown, the turbine ring is attached to the pump impeller's back shroud. The turbine ring is precision cast and the pump impeller is sand cast. The rotating assembly is supported by the stationary axle and thrust plate, employing orifice-compensated, water-lubricated hydrostatic bearings for both radial and axial loads. The hydrostatic bearings are pressure fed from the same high-pressure source that powers the turbine. The bearing inserts were constructed of a carbon-filled Teflon material and performed well over the full-tested flow, head, and temperature ranges.

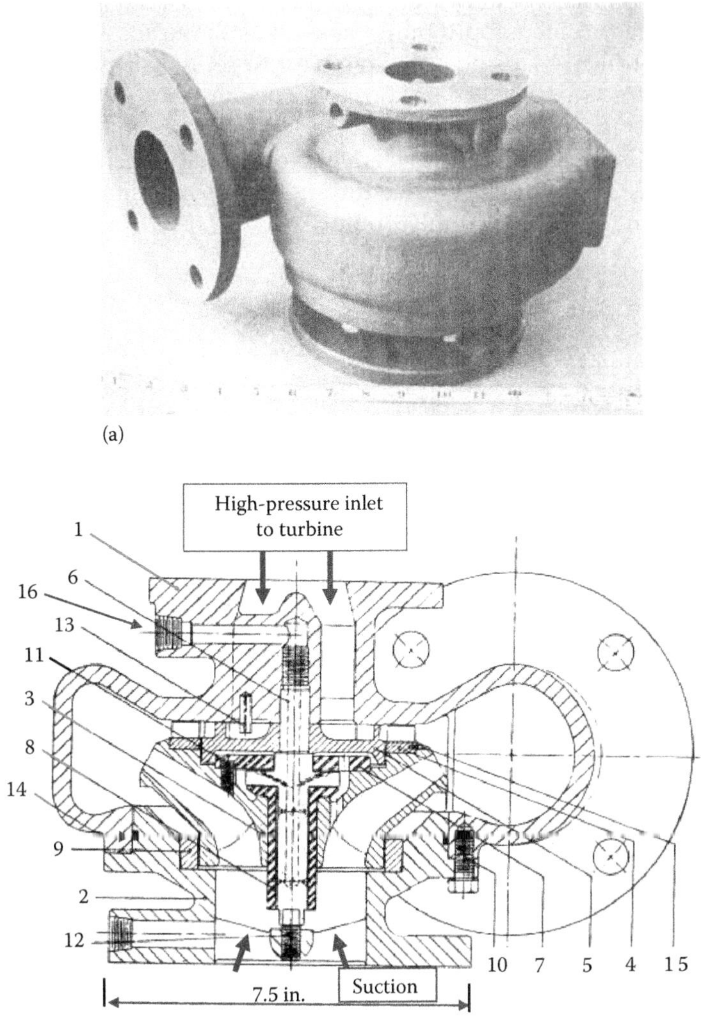

1—Pump casing	4—Turbine ring	7—Rotary thrust plate	10, 11, 12—Screws
2—Suction piece	5—Turbine nozzle	8—Bearing bushing	13, 14, 15—Pin, O-ring
3—Impeller	6—Bearing axle	9—Wear ring	16—Bearing high pressure

(b)

FIGURE 13.2
Combined-impeller turbine-driven pump (a) photo and (b) assembly.

13.2 Test Loop

The laboratory test loop is illustrated in Figure 13.3. As shown, suction pressure to both the turbine supply pump as well as the tested turbine/pump impeller is controlled by using a steam ejector to set the air pressure on the sump's water-free surface. The net positive suction head (NPSH) required of the turbine supply pump was substantially less than that of the tested turbine/pump impeller. Thus, the steam ejector could be used to run cavitation required NPSH tests on the tested turbine/pump impeller without causing cavitation in the turbine supply pump.

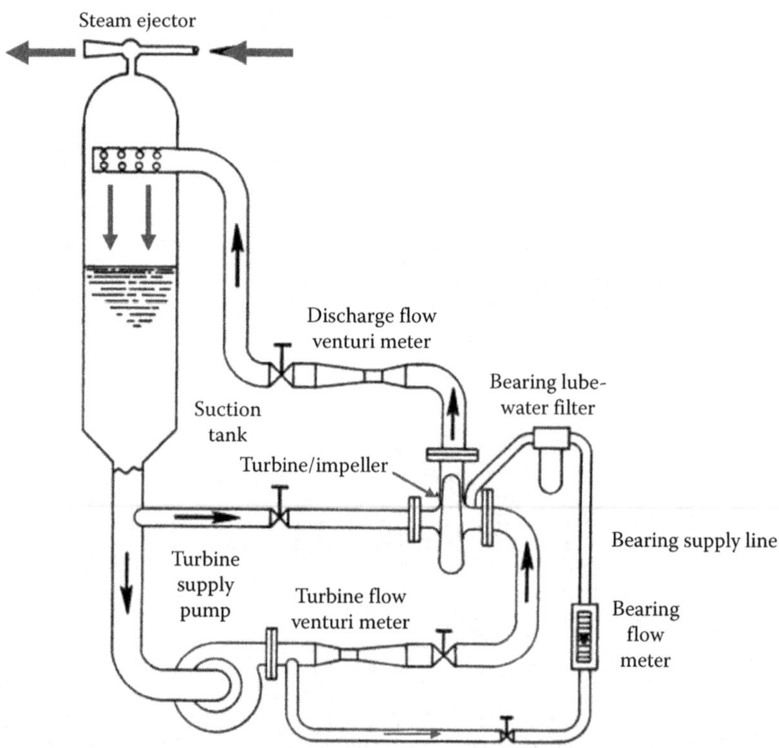

FIGURE 13.3
Schematic of test loop.

13.3 Performance and Endurance Test Results

Performance tests were run on five different variations of the basic unit design. These included the basic design, two with shorter turbine nozzle and blading heights, and two partial-admission nozzle configurations. The peak overall efficiency on the tested design unit was 45%. This compares to the predicted best overall efficiency of 40%, which was computed as the product of the predicted turbine efficiency times the predicted pump efficiency. The predicted NPSH required was 10 feet of water, as compared to the test results which indicated a required NPSH of 15 feet. Not only was the tested unit much smaller in size than an equivalent service jet pump, but the tested unit's 45% best efficiency clearly had a considerable energy conservation superiority over a fixed-nozzle jet pump's 25% efficiency peak and even over a variable-nozzle jet pump's 35%.

The endurance tests indicated that the bearing material (carbon-filled Teflon) and configurations were well suited to this machine. Also, the S-Monel turbine nozzle and blading stood up well against erosion and cavitation, which are inherent in this type of machine. It was determined that reducing turbine nozzle and blade height was a substantially more efficient way to reduce turbine power than using partial emission, that is, selectively blocking some of the turbine nozzle guide-vane passages.

Bibliography

Stepanoff, A. J., *Centrifugal and Axial Flow Pumps*, 2nd edn., Wiley, New York, 462pp., 1957.

14

Water-Lubricated High-Speed Bearings

In the same mid-1960s time period as for the project covered in Chapter 13, the author was quite fortunate to be assigned to a project that began what was to become his specialty, *rotating machinery dynamics, bearings*, and *seals*. A new turbomachinery product development was ongoing for the U.S. Navy to pressurize the air within the ships' double hulls. That was to coat the outer hull's wetted surface with bubbles to prevent sonar detection. Even today, that piece of turbomachinery would be considered highly challenging. It comprised a 42,000 rpm rotor with an overhung centrifugal air compressor impeller at one end and an overhung single-stage impulse steam turbine powering the rotor from the other end. The two journal bearings and two opposed thrust bearings were all hybrid hydrodynamic–hydrostatic fluid-film bearings with water as the lubricant and calculated to be operating in the turbulence regime. At that time, lubrication researchers in the United States and abroad were in the early stages of comprehensively tackling turbulent lubrication films from a theoretical basis, eventually leading to present-day computer codes to calculate turbulent bearing-load capacity and rotor dynamic properties.

In the absence of the not-yet developed computer codes for turbulent fluid-film bearings, the task assigned to the author was to conduct numerous tests using one of the 42,000 rpm prototypes, modified for this research project. The underlying technical approach was to develop an empirically based design methodology by identifying a bearing Reynolds number that would reliably calibrate turbulent bearing-load capacity and stiffness to what laminar flow analyses predicted. This was an approach long used to handle other turbulent flow phenomena like pipe friction. The unit's thrust bearing provided the most convenient way to explore a Reynolds-number-based approach. In optimizing performance and reliability, close attention was focused on not over-designing the bearings. Laminar lubrication bearing design information and guidelines were thus judged to be inadequate for use in these very-high-speed water-lubricated bearings operating in the turbulence regime.

14.1 Product Description and Test Rig

The test program used one of the 42,000 rpm prototypes modified for this project as illustrated in Figure 14.1. The rotor was supported by two bearings,

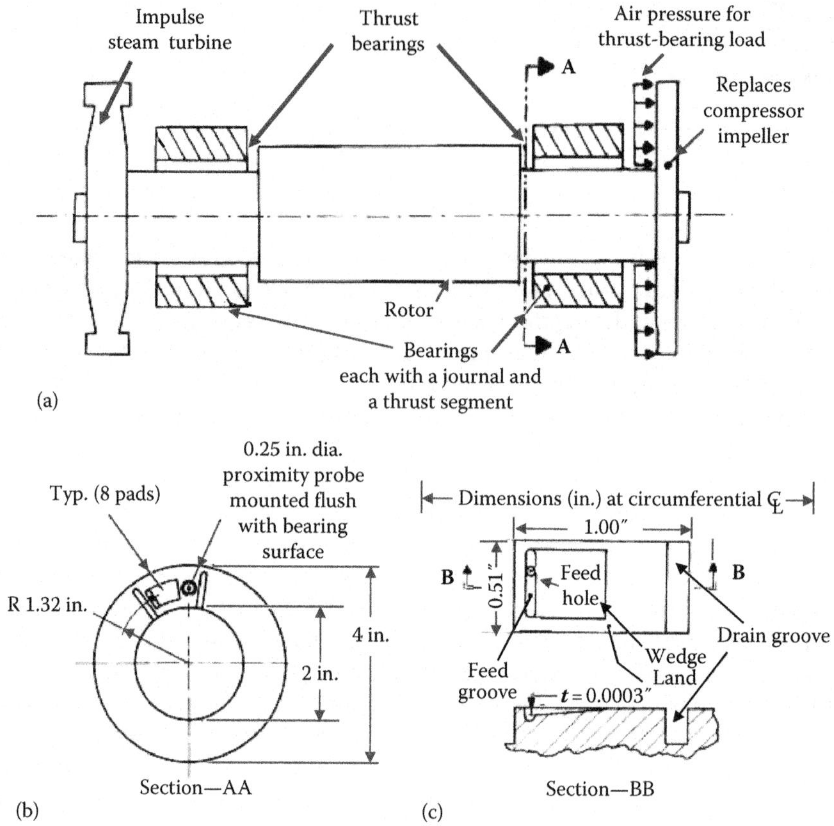

FIGURE 14.1
Test setup for water-lubricated turbulent flow thrust bearing. (a) 50,000 rpm max-speed rotor-bearing unit, (b) thrust bearing, and (c) bearing details.

each having a journal segment and a thrust segment. The centrifugal compressor impeller was replaced by a disk of comparable mass. The disk was circumscribed at its outer diameter with a labyrinth tip seal, so that its trapped internal annular space could be pressurized to apply specified loads upon the compressor-side thrust bearing. The total axial clearance between the shaft thrust shoulders and thrust bearings was sufficiently large (0.030 inch) so that loading from the turbine-side thrust bearing could be neglected. To accurately measure bearing film thickness, a proximity probe was installed as a continuous smooth portion of the compressor-side thrust bearing surface, as illustrated in Figure 14.1. Proximity probes were then a new cutting-edge position measurement development. Not having the actual compressor impeller on the rotor, the prototype impulse turbine was quite capable of driving the rotor to much higher than the rated 42,000 rpm, since it only had

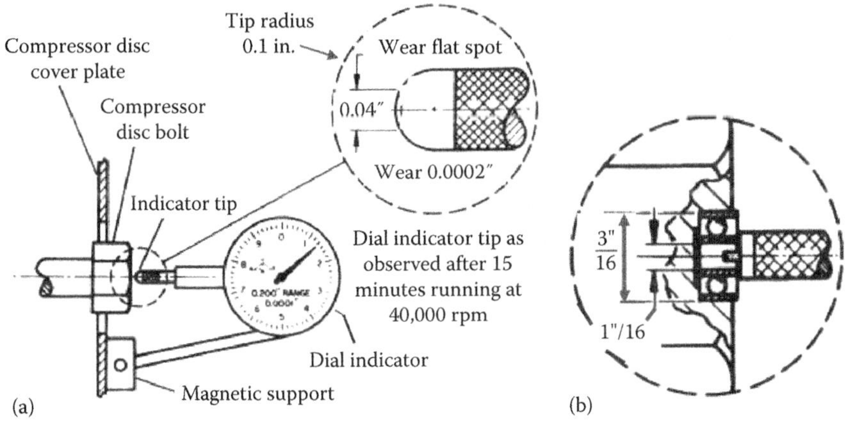

FIGURE 14.2
Dial indicator approaches for the indirect measurement of film thickness: (a) First approach and (b) second approach.

to overcome bearing drag and other secondary losses. Maximum test speed was therefore set by limiting stresses in the rotor. This setup allowed testing over wide ranges of (1) speeds from startup to 50,000 rpm, (2) applied bearing loads, and (3) lube water temperatures.

A significant task of this project was to master using the then-just-emerging new position measurement sensor, *inductance-type noncontacting proximity probes*. These are, of course, now used in industry extensively, for example, in all modern power plants to monitor rotating machinery vibration. In modern rotating machinery research facilities, proximity probes are used not only for rotor vibration measurements but also for bearing film thickness measurements. But that was not the case in the mid-1960s. Our desired accuracy for measuring bearing film thickness was ±0.0001 inch (±0.0025 mm). The task to explore the noncontacting proximity probe was motivated by the lack of measurement accuracy we could achieve with a precision dial indicator. Figure 14.2 illustrates the two approaches first taken using a precision dial indicator. In the first approach, Figure 14.2(a), the tip of the dial indicator wore out too excessively to otherwise work. In the second approach, a very small precision ball bearing was put into the compressor-disk-retaining bolt as shown. Although better than the first approach, it also did not work satisfactorily due to rotor thermal expansion.

The proximity-probe task started by calibrating a probe as illustrated in Figure 14.3. Two important facts quickly emerged that are today well known in the industry. That is, (1) probe calibration is highly repeatable and that (2) the DC output voltage signal is extremely linear with gap over a wide range. As now long well known, this second fact facilitates viewing in real time the enlarged rotor vibration orbit by feeding the DC output signals

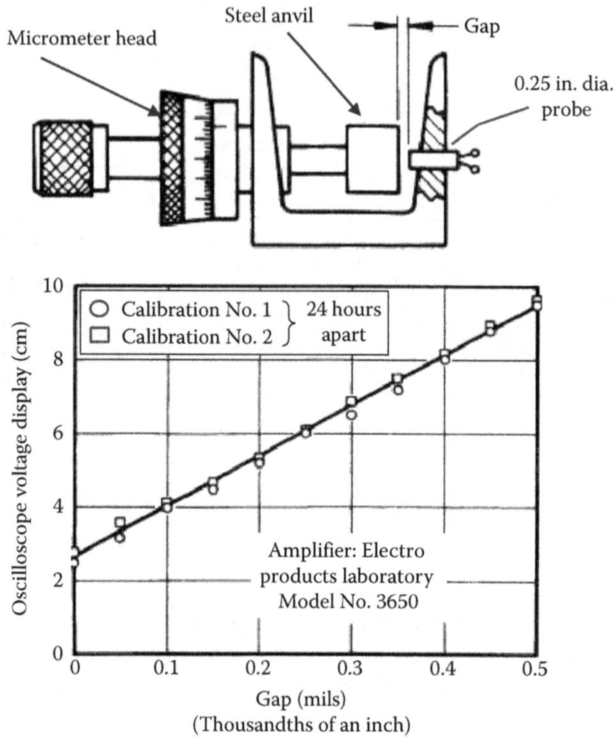

FIGURE 14.3
Example proximity probe calibration (1966).

of two perpendicularly mounted x–y probes into a double-channel oscilloscope. Of course, with modern digital signal processing, the oscilloscope can be replaced by a PC. My late friend Don Bently (1924–2012) was the major pioneer in bringing to industrial applications and researchers worldwide the noncontacting inductance-type proximity probe and supporting instrumentation. Figure 1.10 illustrates the functioning of proximity probes in measuring rotor radial vibration.

14.2 Sample of Test Results

A series of tests were conducted at three lube water temperatures over a range of speed and bearing load. Figure 14.4 shows the results for one of those test series.

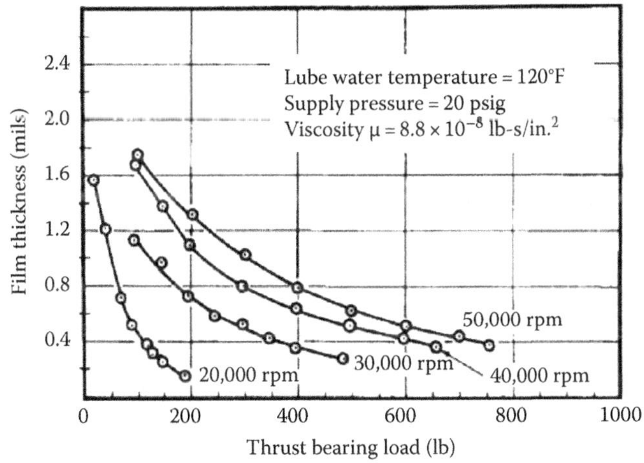

FIGURE 14.4
Example of thrust bearing load capacity test data.

Using that good old fashion log-log paper, *many exercises* were performed with plotting the ratio of measured turbulent load capacity to laminar theory predictions, of course hoping for a straight-line function as follows:

$$\frac{W_T}{W_L} = A\text{Re}^n \tag{14.1}$$

where

W_T is the turbulent load capacity
W_L is the laminar load capacity
Re is the Reynolds number
$\text{Re} \equiv \dfrac{\rho U h}{\mu}$
U is the sliding velocity
h is the film thickness
A and n obtained from plot
ρ is the fluid mass density
μ is the fluid viscosity

The underlying task naturally was to find which bearing film thickness h yielded the best straight-line plot of all tests on log-log paper. "That's engineering." The three bearing film thicknesses tried were the *minimum film thickness* h_2, the *mean film thickness*, and the wedge *inlet film thickness* h_1. The best result was obtained using the inlet film thickness. Referring to Figure 14.1(c), that is given as follows:

$$h_1 = h_2 + t, \quad \text{where } t = 0.0003 \text{ in.} \tag{14.2}$$

That log-log plot is shown in Figure 14.5.

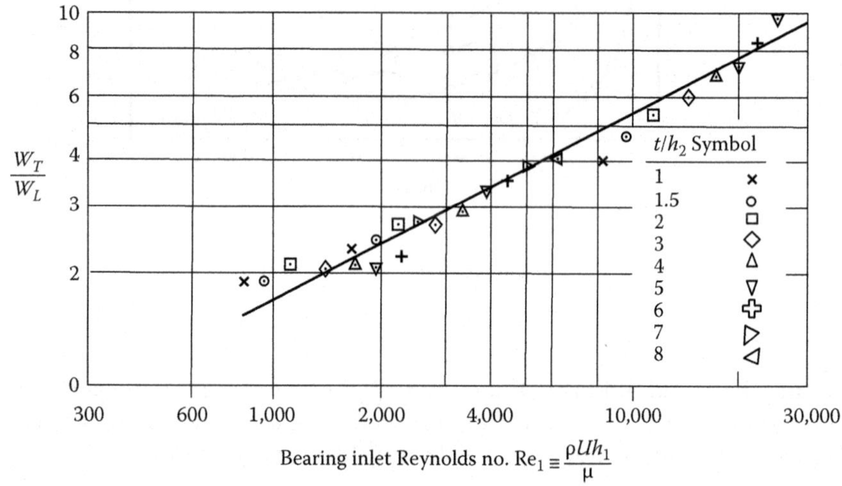

FIGURE 14.5
Load amplification ratio vs inlet Reynolds no.

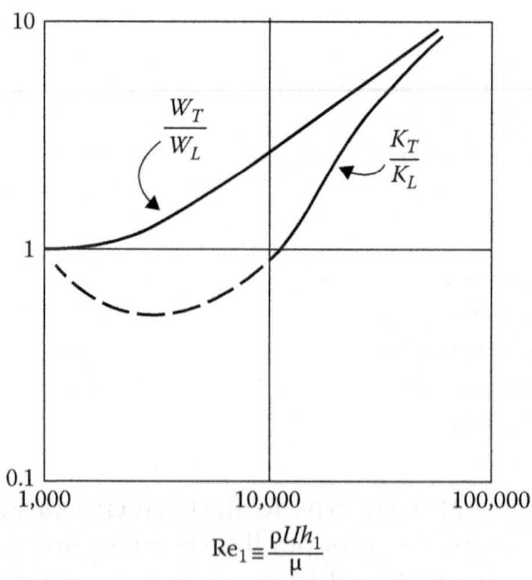

FIGURE 14.6
Comparison between load and stiffness ratios.

The same overall approach was used to correlate *turbulent bearing stiffness* with *laminar theory predictions*. The experimental stiffness was computed by numerically differentiating the measured bearing load test data, Equation 14.3.

$$\frac{K_T}{K_L} < \frac{W_T}{W_L} \quad \text{and} \quad K_T = \frac{dW_T}{dt} \cong \frac{\Delta W_T}{\Delta t} \tag{14.3}$$

It was found that for stiffness, the turbulent-to-laminar K_T/K_L ratio vs inlet Reynolds no. was consistently smaller than the W_T/W_L ratio, Figure 14.6. This made sense since when the load is reduced, the film thickness increases which also increases the Reynolds no. And conversely, when the load is increased, the film thickness reduces, which also reduces the Reynolds no. At lower Reynolds numbers, this effect even yielded turbulent stiffness lower than laminar stiffness. This effect of turbulence on film stiffness was corroborated much later by turbulent lubrication researchers. Mathematically differentiating Equation 14.1 yields the same effect.

15

Aircraft Engine Compressor Blade Tip Rubs

At some point in the evolution of the modern aircraft engine business, the airlines and engine manufacturers devised a win-win arrangement for repairing engines. That is, the airlines now pay the engine manufacturers a specific amount for each operating hour of an engine, sort of like an insurance policy. The engine manufacturers reciprocate by performing engine repairs at no additional charge to the airline. In the "good old days," engine manufacturers had *replacement parts as a high-profit business*, similar still in many other original equipment manufacturer (OEM) replacement parts businesses, most prominently automotive. So now, the engine manufacturer has a higher motivation to design for minimizing the need for replacement parts, especially the costly engine blades. With the resulting added reliability and reduced engine-out time for repairs, the airlines now do not need as many spare engines on ready standby as before, thus reducing their operating costs. This is clearly a win-win arrangement for the engine manufacturers and airlines alike.

Consequently, starting in the late 1990s as part of minimizing engine repair costs, the General Electric Aircraft Engines group in Evendale, Ohio, undertook a long-envisioned research project to study the tip rub dynamics of engine blades. Although nearly every engineering aspect of aircraft engines has at least some safety aspect, this research was primarily motivated by cutting engine repair costs.

15.1 Engine Blade-Tip-on-Shroud Contact in Service

The generic problem of the engine blade-tip rub phenomenon is illustrated in Figure 15.1, albeit with exaggerated deflection. The shroud contact force exerted upon the blade tip comprises radial, tangential, and axial components. The tangential and axial components are essentially from rubbing-contact sliding friction (Turner 2005, Turner et al. 2005). Blade tip contact with the shroud is an unavoidable reality of a jet engine's *normal operation*. The radial clearance at blade tips is desired to be as close to zero as possible during operation to maximize engine efficiency—full consumption being a major operating cost. Tip clearances of new engines are thus typically set at the factory as follows. An engine is assembled cold with blade radial clearances smaller than the operational radial growth of the blades from at-speed

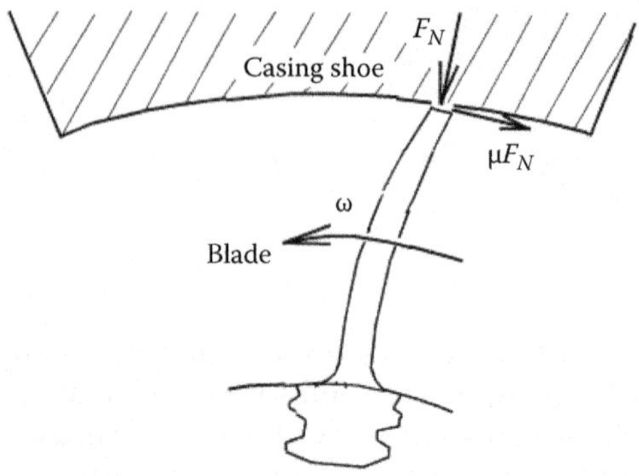

FIGURE 15.1
Single rubbing blade on casing. (From Turner, K. et al., Simulation of engine blade-tip rub induced vibration, *Proceedings, ASME International Gas Turbine Conference*, Reno, NV, June 2005.)

elastic deformation plus thermal expansion. The new engine is then run up to redline speed for a short period, allowing the blades to wear in to their shipped radial dimensions. That produces the tightest possible new-engine blade tip clearances in operation, not achievable only from room temperature manufacturing dimension tolerances.

During normal engine operation, blade-on-shroud contacts are caused by recurring transients imposed upon the engine such as (1) hard landings, (2) rotor gyroscopic effects during abrupt maneuvering, and (3) landing approaches when pilots need to rev the engines back up to boost thrust to correct for undershooting the landing path. In this last one, when first slowing, the engine revs down; the cooling effect on the casing/shrouds causes them to thermally contract. So subsequently when the pilots suddenly rev the engines back up, some interference naturally occurs between blade tips and shrouds. Such normal operational blade-on-shroud contacts can degrade engine performance through the occurrence of high-amplitude rotor vibration and severe blade/seal wear, in the worst case scenario potentially leading to catastrophic engine failure (Padova et al. 2005). Excessive accumulated wear and/or fatigue damage on blade tips necessitate blade replacements, a cost that engine manufacturers, of course, are now highly motivated to minimize.

15.2 Description of Test Rig

A complete unabridged description of the test facility is presented by Padova et al. (2005). It is one of the most challenging test facility developments in which the author has ever participated. The mission of the rig development was to replicate blade-tip-on-shroud rub contacts with an actual engine core compressor bladed disk, and to accurately measure the blade tip force as it travels across the shroud contact arc at speeds of up to 20,000 rpm. At the same time, blades were instrumented with an array of strain guages. This was certainly a tall order and not without some pitfalls along the way to success. The test spindle is shown in Figure 15.2.

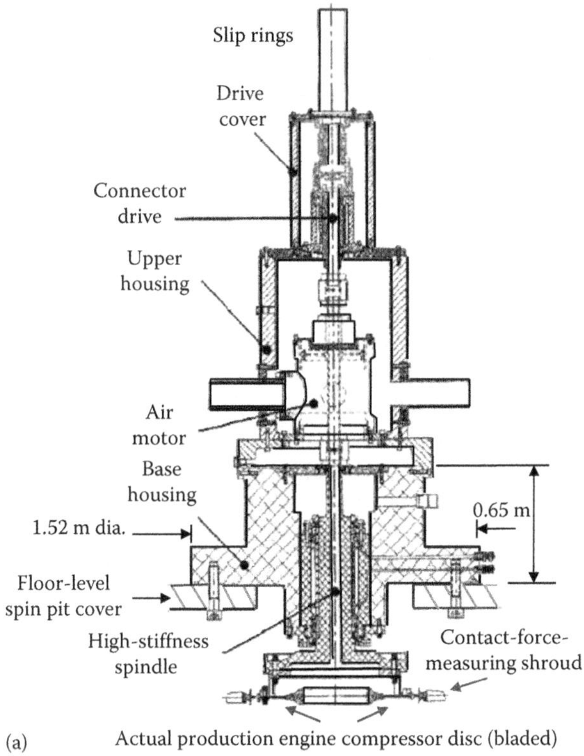

FIGURE 15.2
Vertical test spindle (a) section detail and (b) photo.

An underground containment spin pit houses both the bladed engine compressor disk and the high-stiffness spindle and is spacious enough for technicians to work when setting up tests. As illustrated, the spindle is air-motor driven to speeds of up to 20,000 rpm. Several channels of real-time analog measurement signals are transmitted up to the slip rings through the concentric hole in the spindle illustrated in Figure 15.2(a). These signals are from the multitude of strain guages mounted on selected compressor blade surfaces to identify specific natural-frequency modes of blade vibration and to extract high-cycle fatigue information. The high-stiffness spindle is supported by hybrid ceramic ball bearings to withstand 20,000 rpm operating temperatures for the duration of a single test and to virtually eliminate spindle deflections caused by the imposed blade–shroud contacting. With the high-stiffness spindle, the lowest rotor vibration critical speed determined through analysis was 23,500 rpm, comfortably above the intended maximum 20,000 rpm operating speed.

A fast-acting incursion mechanism had to be devised to produce a quick radial displacement of the shroud into the path of a blade. The conditions of hard landings and abrupt flight maneuvers are reproduced by the partial-arc contacting. With the cam activation mechanism illustrated in Figure 15.3, the incursion mechanism support ring illustrated in Figure 15.4 was rapidly controllable. Its rapid swing action allowed controllable casing sector radial motion to reproduce both a single rub and multiple consecutive rubs of a single blade as well as for adjoining blade clusters. The multiple-consecutive-rubs-type test replicated the influence of a blade's rub-induced vibration upon immediate subsequent rubs, that is, the next revolutions' rubs. This was shown to produce a variety of blade flexing motions appropriately categorized as somewhat *random vibration*.

15.3 Measurement of Blade Tip Contact Force

As illustrated in Figure 15.4(a), three load-measuring units (LMUs) stiffly attach the casing sector to the cam-activated support ring. These three LMUs are extremely stiff and high-precision Kissler™ piezoelectric crystal load cells, each with triaxial xyz force measuring output signals. Thus, there are *nine force signals* to process for the determination of the blade tip instantaneous rub contact force, Figure 15.5. The actual blade tip force and the three triaxial load cells are on opposite sides of the casing sector. The force signals arriving at the load cells are thus only a response to the blade tip force components. Furthermore, the blade is moving tangentially at very high velocity while the load cells are not. Thus, extracting the blade tip instantaneous time-varying force components from the LMU signals presented a unique and considerable challenge, that is, a research project all on its own.

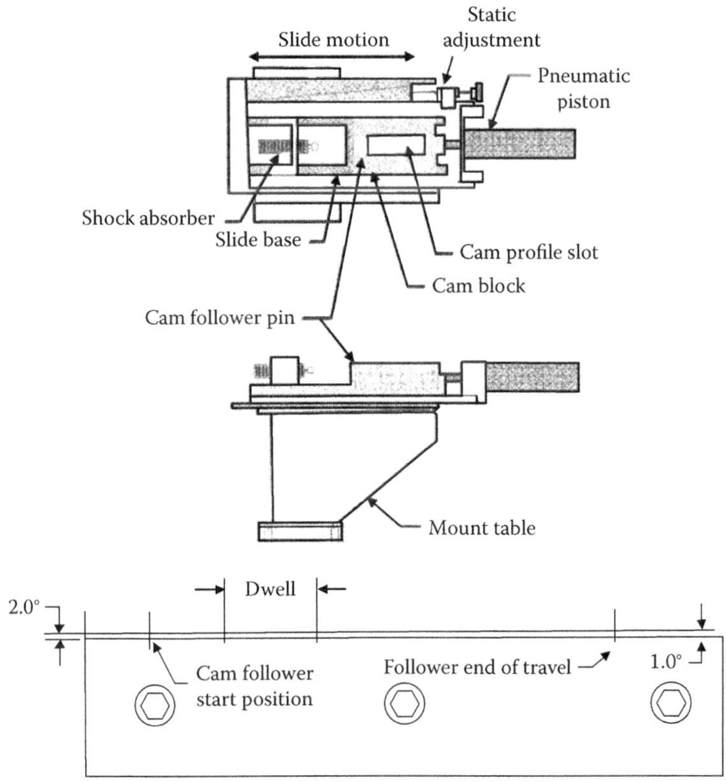

FIGURE 15.3
Incursion mechanism, cam and slide. (From Padova, C. et al., *ASME J. Turbomach.*, 127, 726–735, October 2005.)

A unique *inverse filtering* data reduction software package was developed by Dr. Michael Adams (now a professor and department head at Cleveland State University) to extract blade tip force components from the three tri-axial load cells on the back side of the casing sector. This required meticulous calibration of the load-measuring system by applying vectored impact hammer hits at positions along the sector rub contact surface. The instantaneous tangential angular position of the blade needed to be included in the tip force recovery scheme to properly utilize, with the impact hammer calibration, data taken at specific locations along the casing sector contact surface, Figure 15.5. With blade arrival and exit triggers and with precise rotor speed measurement, the instantaneous blade tangential position was thus known in real time relative to the casing contact sector. Applying the *inverse filtering* method in this application was truly a quantum leap all on its own. Noting from Figure 15.5 in the view shown, the three load cell positions form a triangle, not a straight line that was to equilibrate moments

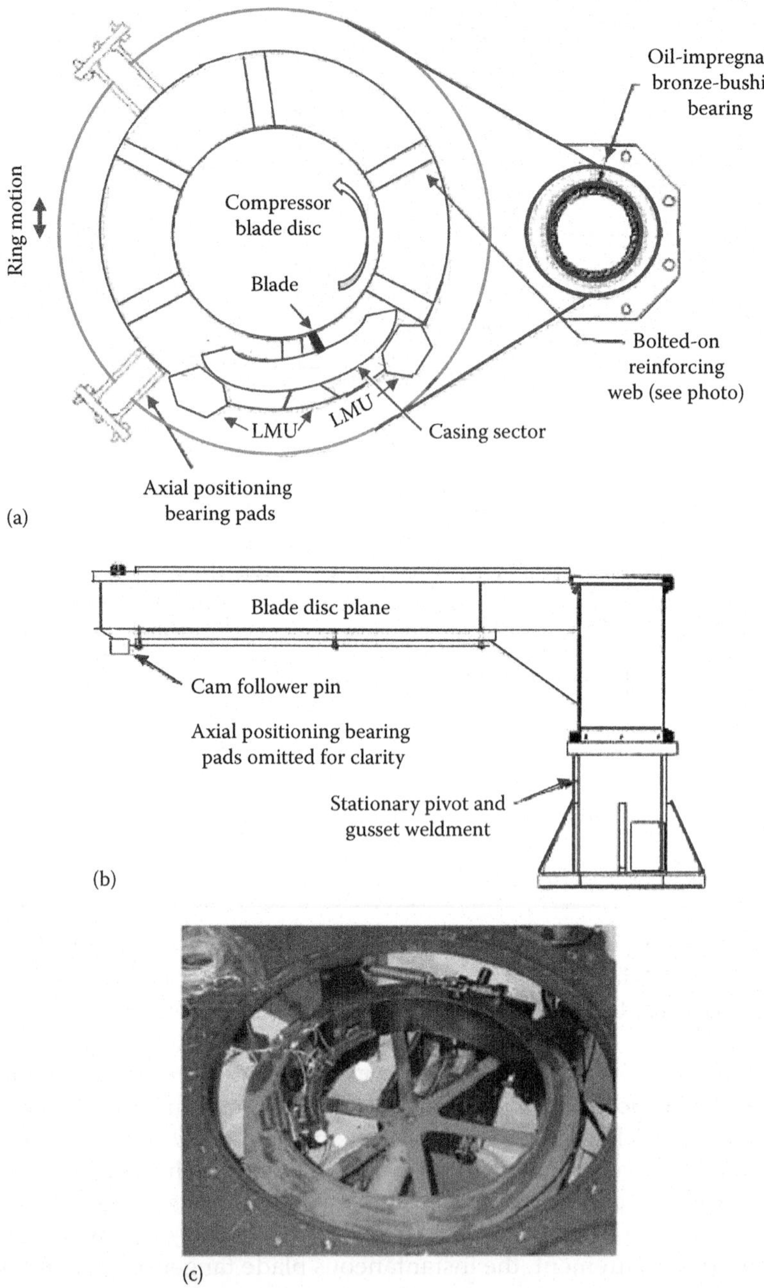

FIGURE 15.4
Incursion mechanism support ring: (a) vertical view, (b) horizontal view, and (c) photo.

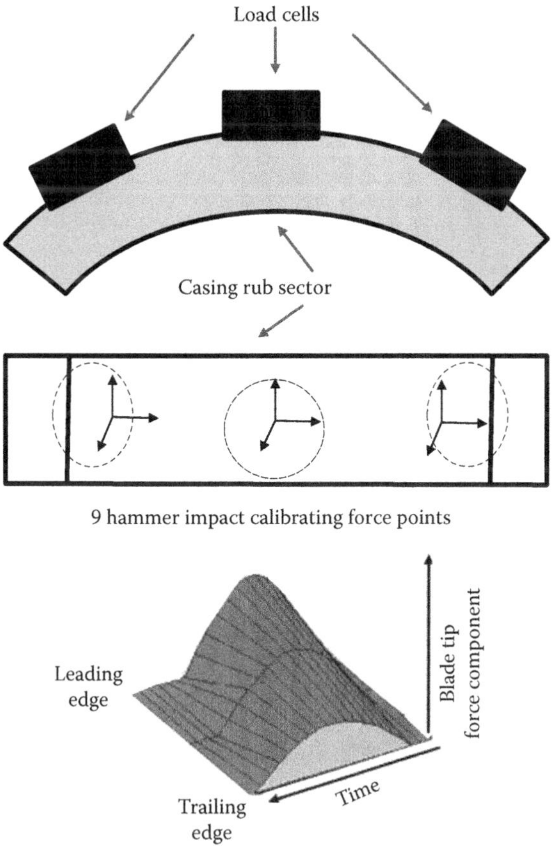

FIGURE 15.5
Casing sector, load cell placements, and force vectors.

imposed upon the casing sector by the contacting blade tip as explained by Padova et al. (2005).

Refining this method required simulating the casing sector + LMU system's structural dynamic response to calibrating the hammer hits using an finite element analysis (FEA) model (ANSYS™). The FEA model included the casing sector assembly, load cells, and incursion mechanism support ring. A sample initial trial comparison between the FEA model and actual hammer hit data is shown in Figure 15.6. The FEA model was iteratively improved, especially LMU stiffness and damping properties, to achieve very good comparisons with actual hammer-hit load cell response data.

As anyone who has ever done real cutting-edge experimental research knows, one always realizes after-the-fact how "it" could have been done

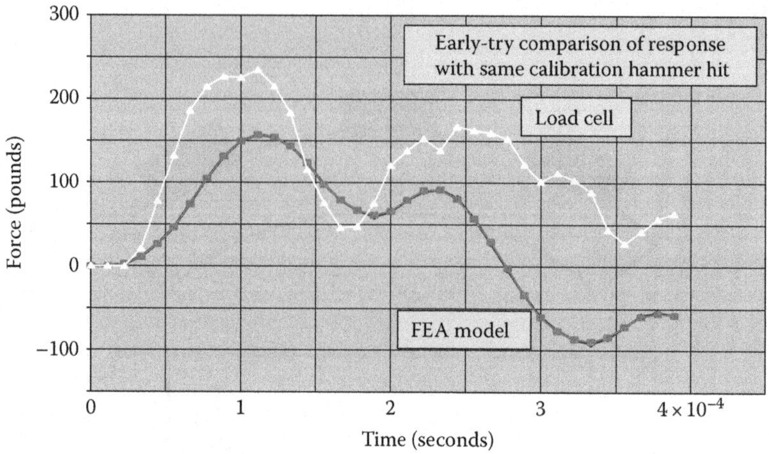

FIGURE 15.6
Sample comparison of a load cell output to FEA model.

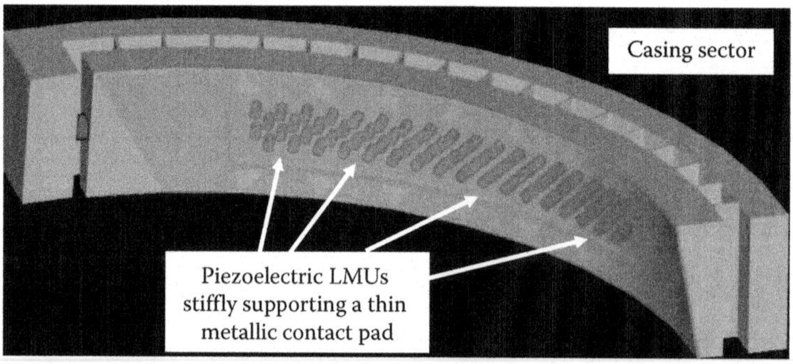

FIGURE 15.7
Multiload cell alternative configuration to measure blade tip forces.

better. Figure 15.7 shows an example of that. Illustrated is the author's proposed alternative approach, which employs 70 smallest-size-available piezoelectric crystal load cells very stiffly supporting a relatively thin metallic arc-of-contact pad. To the author's knowledge, this proposed approach has not yet been tried. Such a multitude of closely spaced load cells near the contact surface would yield a quite rich array of force signals. Processing that near-the-rub-surface array of signals would quite likely yield more accurate and smoothly fitted time-and-position-varying blade tip distributed force scalar components, as ideally pictured in Figure 15.5.

15.4 Companion Research

A companion task for this laboratory research effort on compressor blades was undertaken by the author to develop software to simulate the time-transient nonlinear vibration response of engine blades from operating-speed incursions with the casing (Turner 2005, Turner et al. 2005). The laboratory testing of compressor blade incursions here described led to a follow-on laboratory project to build a new Large Spin Pit Facility to study blade tip rubs for engine fan blades and low-pressure turbine blades. That facility is covered in Chapter 7.

Bibliography

Turner, K., Simulation of engine blade tip-rub induced vibration, MS thesis, Ohio State University, Columbus, OH, 2005.

Turner, K., Adams, M. L., and Dunn, M., Simulation of engine blade tip-rub induced vibration, *Proceedings, ASME International Gas Turbine Conference*. Reno, NV, June 2005.

Padova, C., Barton, B., Dunn, M., Manwaring, M., Young, G., Adams, M. L., and Adams, M. L., Development of an experimental capability to produce controlled blade tip/shroud rubs at engine speed, *ASME Journal of Turbomachinery*, 127, 726–735, Oct. 2005.

16

Centerless Grinder Inside-Out Pivoted-Pad Bearing

Through research conducted at Eaton Corporation (EC) Manufacturing Technologies Center (MTC), Cleveland, OH, research sponsored by Oak Ridge National Laboratories demonstrated that high-speed grinding principles could be successfully applied to the finishing of silicon nitride (ceramic) with considerable improvements in throughput, costs, and quality (Kovach and Malkin 1998). The governing ceramic physical characteristic that yielded this discovery is the increase in ceramic material cutting compliance under very high strain rates. From that discovery, it was immediately realized that exploiting it would require a new generation of high-speed, high-power, centerless grinding wheels that is capable of 7000 rpm and a power output of 50 hp and with very high-stiffness spindles (up to 2×10^6 lb/in., 3.5×10^8 N/m).

To meet these life, stiffness, speed, and power requirements of a next-generation grinding spindle, the author was retained by EC-MTC to devise a spindle bearing system that would meet these new challenges. That effort resulted in a novel inside-out three-pad pivoting-pad oil-fed hydrodynamic journal bearing. The new bearing prototype was built by EC-MTC and tested in the author's rotating machinery research laboratory at Case Western Reserve University (CWRU). The new bearing configuration was an inside-out version of the conventional pivoted-pad journal bearing (PPJB). Two of the three pivoting pads employ fixed-support pivot points while the pivot point of the third pad is supported by a hydraulically actuated radial-position preloading piston. This radial pad actuation provides a real-time controllable preload to the other two pads and thereby controls in real time the overall radial stiffness of the bearing. This spindle bearing stiffness control feature provides the optimization of a grinding cut, whether it's an early deep cut better suited to lower spindle bearing stiffness, or a light shallow finishing close-tolerance cut better suited to higher spindle bearing stiffness. Extensive bearing laboratory test data taken in the author's CWRU laboratory compared favorably with the corresponding predicted theoretical bearing performance (Laurich 1999, Adams and Laurich 2005).

16.1 Generic Centerless Grinding

The fundamental concept of centerless grinding is shown in Figure 16.1. The cylindrical work piece being ground has its tangential velocity imposed upon it by the regulating wheel through frictional traction, with the grinding wheel tangential velocity naturally opposing that of the work piece. The regulating wheel and the grinding wheel are set up with their respective rotational centerlines mutually out-of-parallel by a small angle (e.g., 6°). This provides the means for the regulating wheel to "pull" the work piece axially through the *three converging contact-point lines* (S, R, and G) formed by the work rest blade and the two wheels. As the work piece exits from a centerless grinder, its ground diameter is set as the circle defined by the three work-piece-exit contact points G, R, and S, Figure 16.1. Clearly, the three work-piece-contacting components (two spindles and blade) must have quite high structural stiffness when producing very-high-precision finished ground diameters. For example, IC engine hydraulic valve lifters are mass produced using centerless grinders with a finished diameter and straightness tolerance

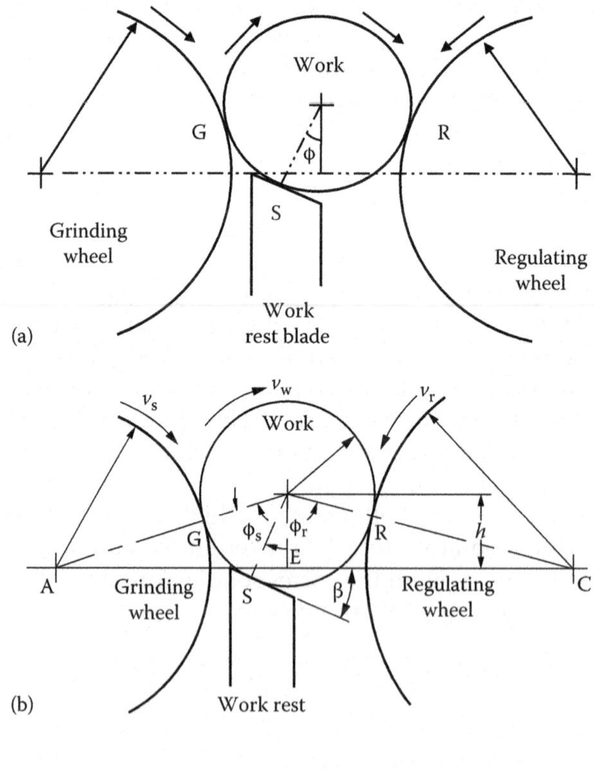

FIGURE 16.1
Generic illustration of centerless grinding: (a) overall view and (b) parameter-detailed view.

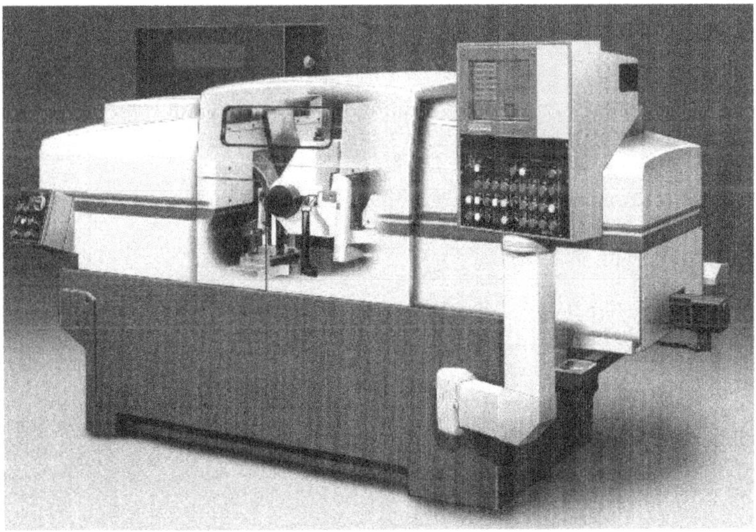

FIGURE 16.2
Modern centerless grinding machine.

of + or –20 millionth of an inch (5.1 ten thousandth of a mm). Figure 16.2 shows a modern centerless grinder.

16.2 Centerless Grinder Wheel with Inside-Out Journal Bearings

The conventional motor driving the grinding wheel is either coaxially coupled to it or connected to it through a belt drive. To achieve the new-generation high-speed, high-power, high-stiffness centerless grinding wheel, the author collaborated with EC-MTC to develop the configuration illustrated in Figure 16.3. This configuration accepts 12–14-inch (30.5–35.6 cm)-diameter and 8–12-inch (20.3–30.5 cm)-wide grinding wheel sizes. It accommodates both *in-feed* and *through-feed* modes. It integrates the motor within the grinding wheel and between the spindle bearings to minimize space requirements (Kovach and Laurich 1998).

At the heart of any machine tool spindle are the bearings. Conventional grinder spindles generally use angular-contact ball bearings, which have virtually no vibration-damping capacity but can be preloaded to give adequate stiffness, albeit at the cost of shortened bearing life. Alternatively, conventional hydrodynamic fluid-film cylindrical sleeve journal bearings lack spindle centerline position accuracy due to bearing radial clearance. On the

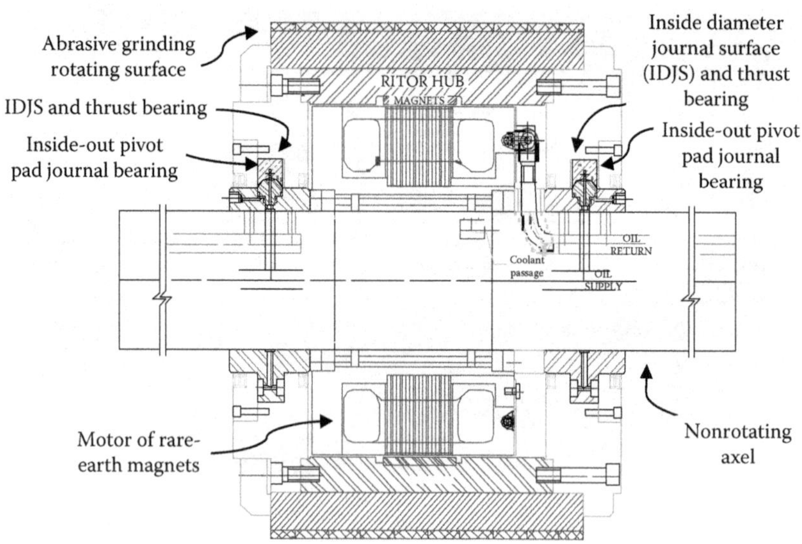

FIGURE 16.3
Centerless grinder spindle with internal 50 hp motor.

other hand, hydrodynamic fluid-film PPJBs can be configured to produce position accuracy, high stiffness, and vibration damping. This is because of the option for a PPJB to be radially preloaded by imposing a pivot clearance smaller than the ground clearance (Adams 2010). Any hydrodynamic bearing, of course, requires a constant supply of lubricant, since a hydrodynamic bearing acts as a viscous pump.

The novel spindle configuration in Figure 16.3 was designed to accommodate the requirements to meet the life, stiffness, speed, and power required to high-speed grind ceramic cylinders. The grinder spindle motor is designed of *rare-earth magnets* as indicated in Figure 16.3.

Two inside-out PPJBs, Figure 16.4(a), were designed to provide the required radial support stiffness of the grinder wheel. As shown, the design of the bearings additionally included an annular axial-face hydrodynamic 3-fixed-pad slider thrust bearing. With the spindle thus axially captured between these two thrust bearings, each axially preloading the other, high-stiffness spindle axial support was achieved as well. The inside-out 3-pad pivoted-pad hydrodynamic journal bearing was configured so that the ID-surface journal rotates outside the bearing pads, which are mounted on a fixed non-rotating axle, as shown in Figure 16.3. To provide automatic pivoted-pad axial 3-axis self-alignment and very high pivot stiffness that is virtually rigid compared to the bearing fluid films, a ball-and-socket pad pivot configuration was employed, Figure 16.4(b). In operation, two of the three pivoting pads support the grinding forces, rotor weight, and radial preload imparted

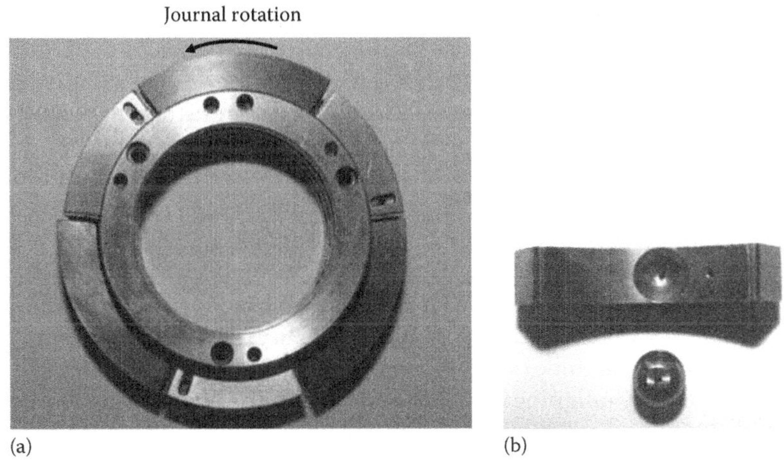

FIGURE 16.4
Bearing prototype tested: (a) complete and (b) single pad.

by the third (top) pad, which is force-controlled by a hydraulically actuated piston, Figure 16.5.

Since three pivot points define a circle, employing the 3-pad configuration allows the bearing preload to be set by the radial position adjustment of only one pad pivot point relative to the other two pad pivot points. By hydraulically actuating the preload pad, the bearing preload and thereby

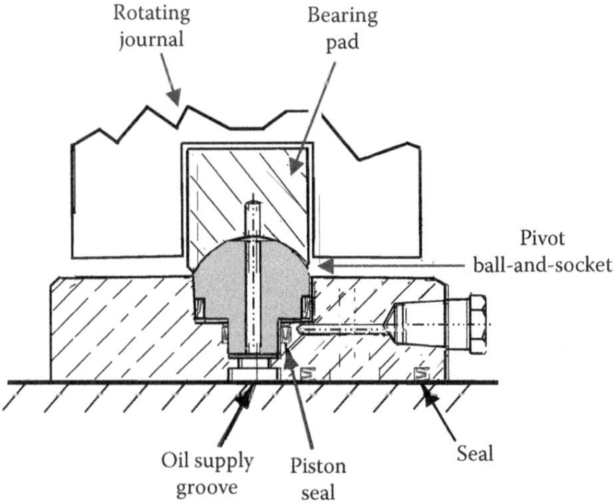

FIGURE 16.5
Preloading pad hydraulic actuation piston arrangement.

the net radial fluid-film stiffness of the bearings are accurately controllable in real time, since stiffness increases with preload (Adams 2010). This adds considerably to operational versatility, providing less stiff spindle bearings for initial rough grinding and very-high-stiffness spindle bearings for precision finish grinding. The three pivoting pads are made of copper to better dissipate heat from viscous energy losses in the hydrodynamic fluid films.

The development of this bearing was guided by extensive analyses of all the important standard fluid-film bearing operating parameters, including lubricant flow requirements, minimum film thicknesses, film operating temperatures, and film power dissipation, all as functions of preload and a spindle speed of up to 7000 rpm (Laurich 1999). Regarding minimum film thicknesses, since the three bearing pads are fully free to pivot, they are automatically freely self-aligning to the journal and thus able to reliably operate with exceptionally small minimum film thicknesses out to eccentricity ratios $\epsilon = e/C$ of 0.99 proven in a high-pressure sliding-vane pump (Adams 1971). Clearly, very fine lubricant oil filtration is a must.

16.3 Bearing Laboratory Testing

The test rig used in this bearing development is covered in Section 1.1. The fixture for applying a controlled static load on the bearing while insuring no journal-to-bearing misalignment is the same as explained with the illustration in Figure 1.8 for controlled journal-on-bearing impact tests (with or without misalignment) and pictured here for these tests, Figure 16.6. The maximum laboratory test speed was 4000 rpm, set by the laboratory test rig.

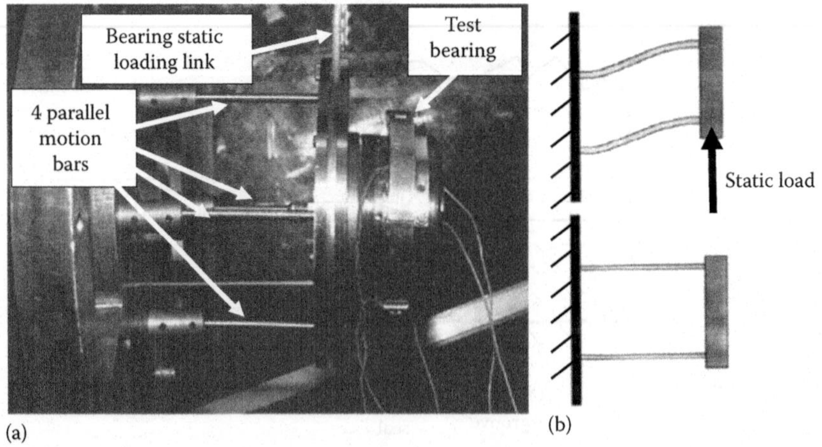

FIGURE 16.6
Static bearing load application arrangement: (a) photo and (b) concept.

TABLE 16.1

Predicted and Measured Bearing Stiffness, 2000–3000 rpm

		Stiffness		
Speed (rpm)	**Actuator Force (lb)**	**Measure (lb/in.)**	**Predicted (lb/in.)**	**Error (%)**
2000	233.3	223,077	277,792	19.7
2250	207.4	242,857	230,059	5.6
2500	219.7	275,610	239,888	14.9
2750	213.3	303,846	305,781	0.6
3000	281.9	329,167	321,592	2.4

The large volume of test data from this work is contained in the thesis by Laurich (1999). The most important test results and comparisons with predicted performance are tabulated here in Table 16.1. It is important to point out that experimental stiffness under static load must be determined by taking the slope of the experimental static load vs. bearing radial displacement. Using predicted bearing stiffness at speeds of up to 7000 rpm and higher static loads, the predicted maximum bearing stiffness was computed as follows:

$$\text{Maximum radial stiffness per bearing: } 1\times 10^{6}\ \text{lb/in.}\left(1.75\times 10^{8}\ \text{N/m}\right)$$

The good comparisons shown in Table 16.1 for 2000–3000 rpm were quite reassuring, since the new bearing configuration could provide the maximum spindle stiffness values up to 7000 rpm judged necessary to fully exploit the demonstrated high-speed ceramic grinding compliance (Kovach and Malkin 1998)

Bibliography

Adams, M. L., Designing high-load journal bearings, *Machine Design*, Penton Publishing Co., Cleveland, OH, Jan. 7, 1971.

Adams, M. L., *Rotating Machinery Vibration: From Analysis to Troubleshooting*, 2nd edn., Taylor & Francis, CRC Press, Boca Raton, FL, 465pp., 2010.

Adams, M. L. and Laurich, M. A., Design, analysis and testing of an inside-out pivoted-pad journal bearing with real-time controllable preload stiffness, *Proceedings, ASME World Tribology Conference*, Washington, DC, Sept. 2005.

Kovach, J. A. and Laurich, M. L., Next generation grinding spindle for cost-effective manufacture of advanced ceramic components, Final Report, Oak Ridge National Laboratory, Oak Ridge, TN, 1998.

Kovach, J. A. and Malkin, S., High-speed, low-damage grinding of advanced ceramics phase-2, Final Report, Oak Ridge National Laboratory, Oak Ridge, TN, 1998.

Laurich, M. A., Inside-out pivoted-pad centerless-grinder journal bearing with controllable preload and stiffness, MS thesis, Case Western Reserve University, Cleveland, OH, 1999.

17

MIT Gas Turbine Lab

The MIT Gas Turbine Lab (GTL) has had a worldwide reputation for research and teaching at the forefront of gas turbine technology for more than 60 years. GTL's mission is to advance the state of the art in fluid machinery for power and propulsion. The GTL archival and published material for this chapter was provided by Dr. Zoltan Spakovszky, MIT Professor of Aeronautics and Astronautics and Director of MIT's GTL. The author gratefully acknowledges the background summaries provided by Dr. Spakovszky on reasons for and important results from the specific GTL experimental facilities described herein.

17.1 Brief History and Background

The MIT Gas Turbine Lab was conceived not long after the first jet engines were successfully run. Shortly after the end of WWII, Professor J.C. Hunsaker [*US aviation pioneer, member of National Advisory Committee on Aeronautics, first MIT Ph.D. in aeronautics (1916)*] brought together a group of American companies that donated funds for the construction of a laboratory devoted to jet propulsion. A plaque still hangs in the laboratory to commemorate those organizations as follows:

Curtiss Wright
General Electric
General Machinery
United Aircraft
Westinghouse
U.S. Navy

Professor E. S. Taylor, first director of the gas turbine lab, opened the GTL on October 7, 1947. Today, the MIT GTL maintains strong ties with industry and government research in the areas of propulsion and turbomachinery technology, as well as with other academic institutions in the United States and abroad who are leaders in the field.

The research at the GTL is focused on advanced propulsion and energy systems and turbomachinery. Activities have included computational, theoretical, and experimental studies of (1) loss mechanisms and unsteady flows in turbomachines, (2) compression system stability and active control, (3) heat transfer in turbine blading, (4) gas turbine engine noise reduction and aero-acoustics, (5) pollutant emissions and community noise, (6) microelectromechanical systems (MEMS)-based high-power-density engines, and (7) multiphase and nonideal fluid machinery design such as cavitating turbopump inducers and supercritical carbon dioxide compressors.

Examples of past research achievements include the first implementation of a three-dimensional computation of the flow in a transonic compressor and the concept of blowdown testing of transonic compressors and turbines, enabling these machines to be used for university-scale experiments. Other examples are the work on turbomachine instabilities and *smart engines* and *micro engines* involving extensive collaboration with the MIT Department of Electrical Engineering and Computer Science. The *Silent Aircraft Initiative* was a collaborative project with Cambridge University, Boeing, Rolls Royce, and other industrial partners to dramatically reduce aircraft noise below the background noise level in well-populated areas.

More recently, research sponsored under the NASA N+3 program in collaboration with Aurora Flight Sciences and Pratt & Whitney, the GTL developed an advanced commercial aircraft configuration with a boundary layer ingesting, embedded propulsion system dubbed the *double bubble D8* concept. In summary, the GTL participates in research topics related to short-, mid-, and long-term problems and maintains strong ties with industrial and governmental research in the areas of propulsion and turbomachinery technology, as well as with other academic institutions who are leaders in this field.

17.2 DeLaval Subsonic and Supersonic Wind Tunnel and Air System

Central to the GTL is the DeLaval wind tunnel and air system, which consists of a horizontally arranged subsonic loop and a vertical supersonic loop conceived in 1946 by Prof. E. S. Taylor, the first director of the laboratory, Figures 17.1 and 17.2. The wind tunnel air system is supplied by a 1.0 MW 5-stage centrifugal compressor. From 1947 to 2015, the compressor was powered by a 1500 hp DC motor and electrical drive system, which was acquired by MIT in 1947 from the U.S. Navy—salvaged from the decommissioned USS Halibut SS-232—a WWII diesel-electric submarine.

FIGURE 17.1
Archival image of the DeLaval wind tunnel, circa 1947.

FIGURE 17.2
Original control tower for the DeLaval wind tunnel.

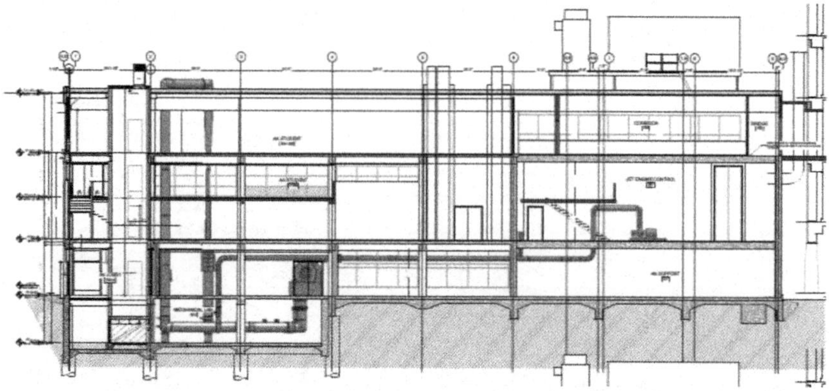

FIGURE 17.3
Upgraded MIT GTL facilities to be completed in 2017.

The subsonic tunnel can be operated as a closed- or open-loop system. The test section is typically 1 × 1 ft with a Mach number range of up to 0.6 and Reynolds numbers of order 10^6. The supersonic tunnel is a closed loop with an 8 × 8 in. test section in a 76 × 20 × 8 in. compartment. The Mach number range is up to 2.0, with total pressure and total temperature adjustable independently via a steam ejector vacuum system with a low-flow compressor and optical instrumentation access. In the 1950s, Prof. A. H. Shapiro carried out seminal compressible flow experiments in this wind tunnel, which contributed to his textbook on compressible flow (1983).

The DeLaval tunnel is also used as an air source and provides flow for various other experiments in the GTL such as, for example, the transonic swirling flow rig used for testing centrifugal compressor vaned diffusers and inlet distortion testing of serpentine inlets for UAVs. A major renovation of the building began in 2015 and will end in 2017, Figure 17.3. The DeLaval tunnel will remain close to its original configuration with the supersonic loop still vertical.

17.3 Rotordynamic Test Rigs

In combustion gas and steam turbines, there is a tangential destabilizing force effect that originates in turbine stages which, in the absence of adequate rotor vibration damping, can produce large-amplitude subsynchronous forward whirling rotor vibration quite similar to the *oil whip phenomenon* initiated by too lightly loaded journal bearings (Adams 2010). Section 4.4 presents the nature of this force as attributed to its first researchers, Thomas (1958) on steam turbines and Alford (1965) on gas turbines. Figure 4.6 illustrates the resultant sum of all the tangential blade forces upon a turbine stage which combine to produce a net destabilizing force that is perpendicular to the rotor radial eccentricity and in the corotational direction. Important subsequent work on this topic by Thomas and his doctoral students at the Technical University Munich (TUM) is documented by Thomas et al. (1976), Urlichs (1975, 1976), and Wohlrab (1975).

This destabilizing force can excite the lowest-damped corotational (forward whirling) rotor vibration orbit. If the available damping—mainly from the journal bearings—is insufficient to control this negative damping destabilizing force, then a large-amplitude forward whirling vibration similar to oil whip will occur. In large high-pressure steam turbines, the lowest-damped forward-whirl mode's frequency is typically near half the running speed frequency, that is, near 30 Hz on a 60 Hz U.S. 2-pole generator. The transition from the incipient instability to the large-amplitude steady-state nonlinear limit cycle is illustrated in Figure 4.5, being limited by the journal bearing radial clearances. This vibration in steam turbines is called *steam whirl*. The author (2010) explains how this effect is now modeled within modern rotor vibration design analyses. Shown here is a troubleshooting case study demonstrating how a cascade of axial direction stationary guide vanes (called *swirl brakes*), just upstream of the tip seal annulus, solved a steam whirl rotor vibration problem, Figure 17.4. The field fix allowed the Unit-1 Brown-Boveri 1300 MW cross-compound steam turbine at TVA's Cumberland plant to operate at full load after previously being restricted to 1100 MW to avoid intolerably high steam-whirl-induced rotor vibration.

MIT GTL experimental work to confirm and extend that of Thomas and Alford is given by Martinez-Sanchez et al. (1995) who presented their experimental results on the magnitude, origin, and parametric variations of destabilizing rotor dynamic forces that arise in high-power turbines due to blade-tip leakage effects. Their tests included five different unshrouded turbine configurations and one configuration shrouded with a labyrinth seal, all tested with static radial offsets of the turbine shaft. The test rig is illustrated in Figure 17.5. The forces along and perpendicular to the offset were measured directly with a dynamometer and were also inferred from velocity triangles and pressure distributions obtained from detailed flow field interrogations.

This work confirmed the existence of the destabilizing force insights proposed by Thomas and Alford. The general scaling and order of magnitude are also consistent with those insights. They found that for shrouded turbines, the upstream and downstream nonuniformities caused by flow redistribution increase the magnitude of the forces by increasing seal pressure nonuniformity. This conclusion appears to be consistent with the more specific insights of Thomas et al. (1976) and Wright (1983) regarding the pre-swirl of circumferential flow and resulting shear stress within the annulus between two adjacent labyrinth tip seals, Figure 4.6.

Compressor whirl research at GTL conducted in collaboration with GE Aircraft Engines Group are documented by Storace et al. (2001), Ehrich et al. (2001), and Spakovszky (2000). These collaborations combined experimental

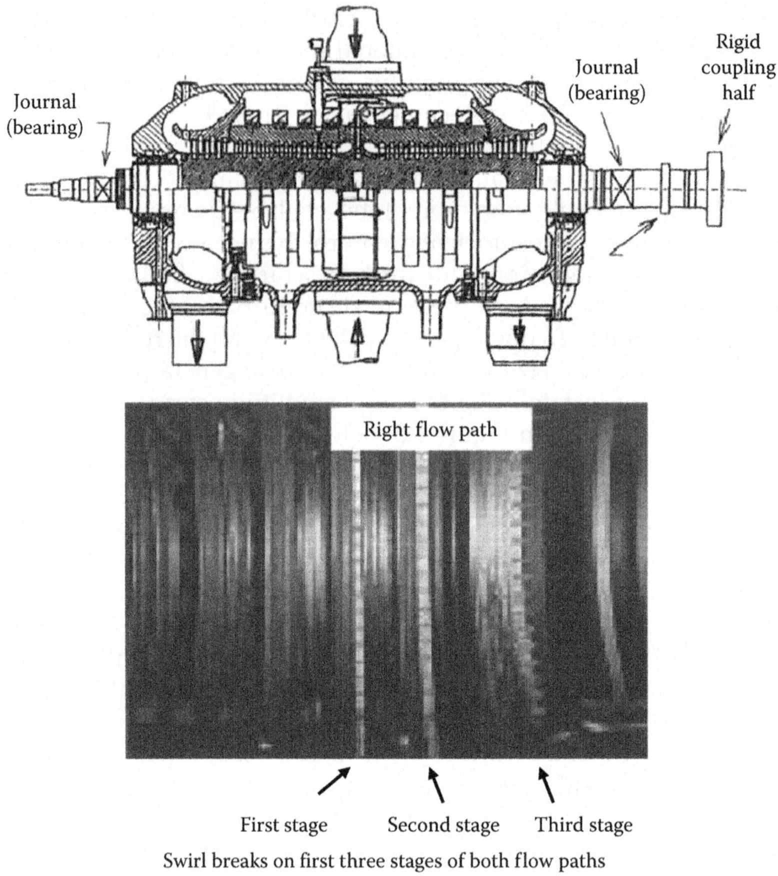

FIGURE 17.4
Double-flow high-pressure steam turbine at TVA Cumberland plant.

and analytical work that resolved a long-standing disparity in findings between whirl direction in turbines vs. compressors. Investigations were focused both on rotor tip clearance and stator shroud-seal clearance, with the rotor systematically offset within the casing. Rotor pressure and suction sides were instrumented with high-response pressure transducers to measure unsteady blade surface pressure from which whirl-inducing force can be determined, Figure 17.6.

Analytical modeling estimated aerodynamic forces in axial flow compressors due to asymmetric tip-clearance (Spakovszky 2000). The model captured effects of tip-clearance-induced distortion from forced shaft whirl, unsteady momentum-induced tangential blade forces, and pressure-induced forces on the spool. Criteria were established to determine conditions and direction for whirl onset based on the blade loading indicator. The blade loading indicator depends only on stage geometry and mean flow coefficient and determines

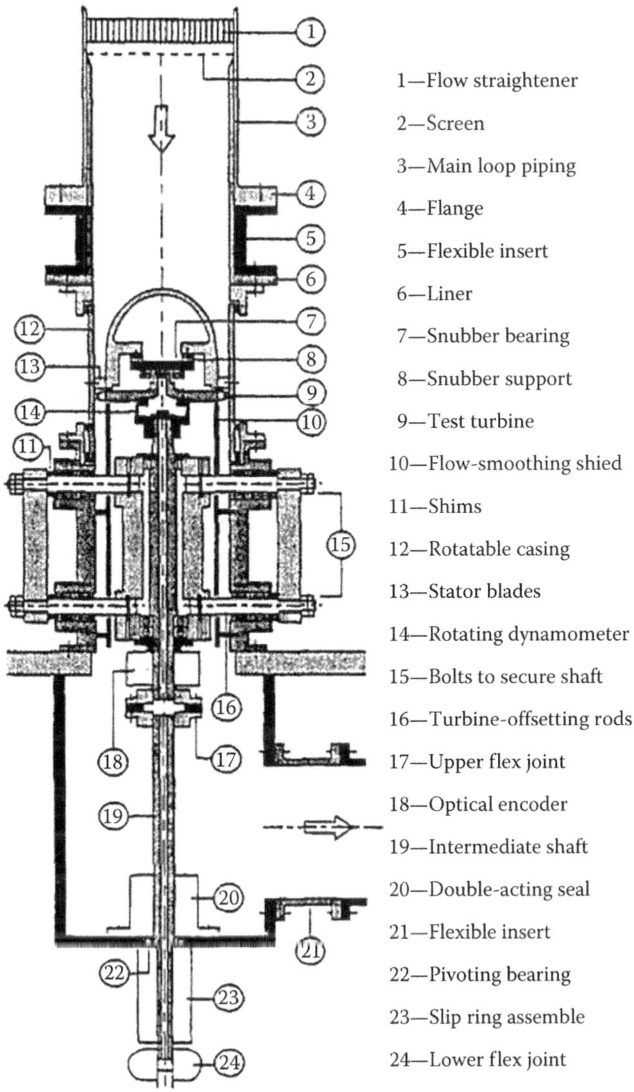

FIGURE 17.5
Schematic of MIT Alford force turbine test rig.

the direction of whirl tendency due to tangential blade loading forces in both compressors and turbines. All findings were suitable for incorporation into an overall dynamic system analysis and integration into existing engine design tools.

Millsaps and Martinez-Sanchez (1994) simultaneously imposed spinning and orbital rotor motion within a single gland labyrinth radial seal. They showed that the cross-coupled destabilizing forces are governed by two key

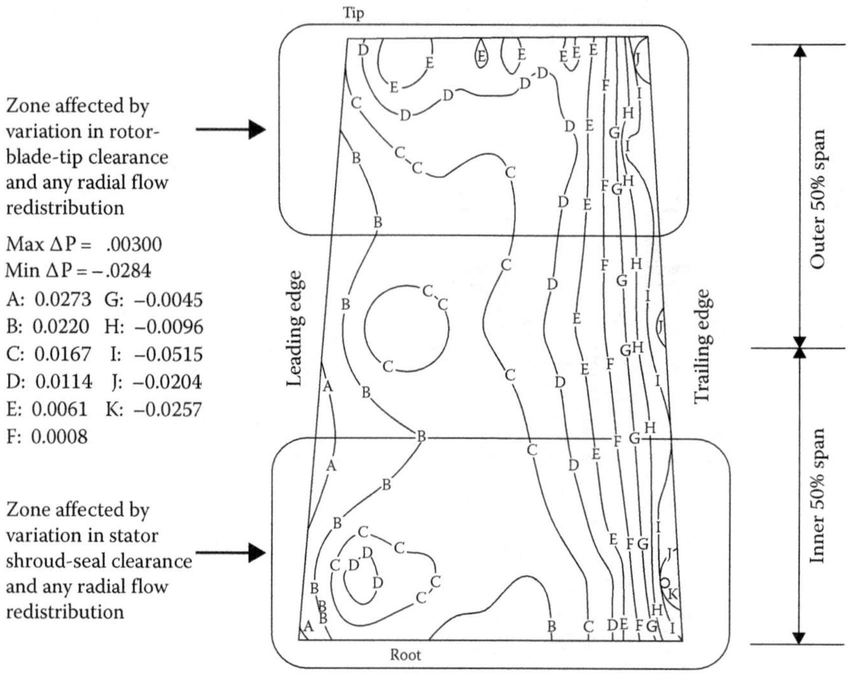

FIGURE 17.6
Measured unsteady blade surface pressure at minimum clearance condition—GE LSRC compressor. (From Ehrich, F.F. et al., *ASME J. Turbomach.*, 123, 446–452, 2001.)

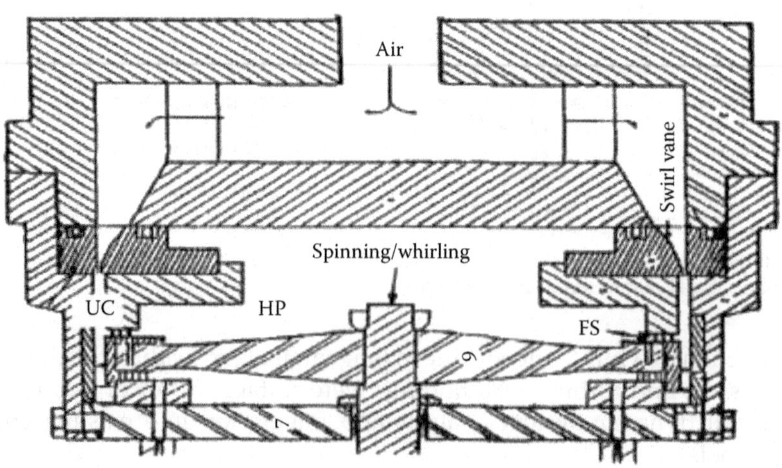

FIGURE 17.7
Labyrinth seal tester with upstream swirl cavity and center hub plenum.

mechanisms: first, an inviscid ideal flow component due to swirling flow contributes to direct damping and second, a viscous component contributes to cross-coupled stiffness and direct damping, deduced from measurements of cross-coupled stiffness. Their test apparatus is illustrated in Figure 17.7.

17.4 Blowdown Testing of Transonic Compressors and Turbines

Kerrebrock (1972) utilized a blowdown compressor facility for the low-cost aerodynamic testing of full-scale compressor stages. They achieved test conditions with both correct Mach and Reynolds numbers while avoiding the high cost of full-scale high-speed compressor testing with large power requirements and high mechanical stresses. The objective of short-duration blowdown testing is to enable fundamental studies of compressor aerodynamics, both steady and unsteady, at low cost, matching Reynolds and Mach numbers to those of full-scale production compressors. Since the time scale of the flow phenomena of interest is less than a rotation period of the compressor, the relevant aerodynamic data can be acquired in a test lasting only a few rotational periods. Furthermore, the use of a high-molecular-weight (low-sound-speed) gas allows testing at a high Mach number without serious stress issues, and the elimination of high-power drive systems enables large-diameter tests at a reasonable cost. Large size combined with low-sound-speed gas reduces blade passing frequency to levels that can be resolved with common pressure transducers. The more or less free choice of the test gas also creates opportunities for optical diagnostics, which are otherwise not feasible.

The compressor rig, Figure 17.8, consists of a supply tank, a fast-acting valve, and an instrumented test section with a compressors stage and dump tank. The gas supply tank behaves like a stagnation plenum, not like a shock tube driver. This requires test time to be large compared to acoustic time of the supply tank. As supply gas expands, pressure and temperature decrease but the rotor slows as work is done, but rotor tangential Mach number is kept nearly constant by the proper choice of sufficiently large rotor inertia. If the discharge orifice remains choked, the axial Mach number is constant. Initial dump tank pressure is set by power requirements for bringing the rotor to speed. The vacuum requirement for the dump tank is modest. The instrumented fan test section is shown in Figure 17.9.

The GTL blowdown turbine facility follows similar ideas with the turbine test section illustrated in Figure 17.10 (Epstein et al. 1985, Abhari and Epstein 1994). The turbine tends to speed up as work is done on it, so a magnetic eddy current brake is employed, illustrated in Epstein et al. The facility is capable of testing 0.5-m diameter film-cooled high-work aircraft turbine stages under actual engine conditions. Simulation capability includes turbine inlet pressures of up to 40 atm and inlet temperatures of up to 2500 K. The focus of this

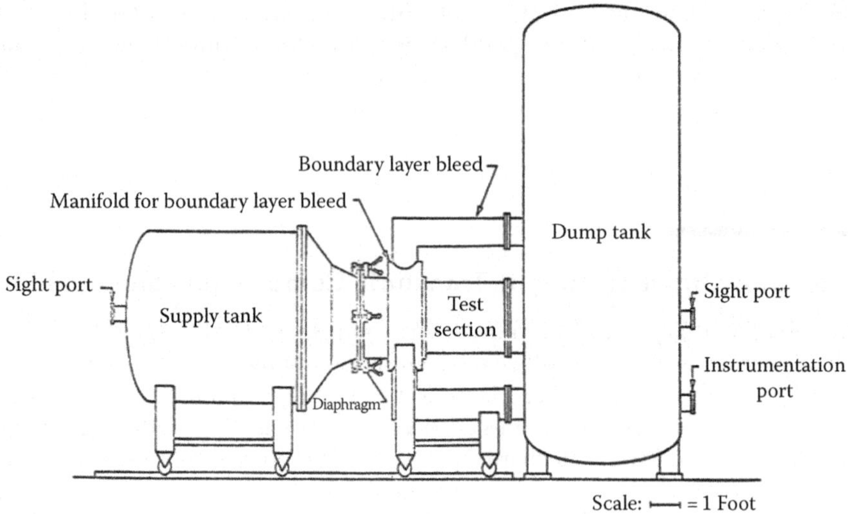

FIGURE 17.8
Blowdown compressor facility housing a 23.25-inch-diameter rotor.

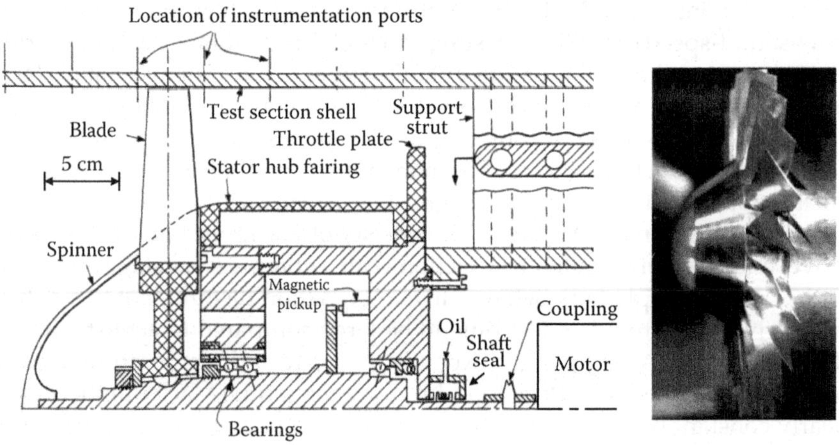

FIGURE 17.9
Fan test section in a blowdown facility.

research was on the exploration of unsteady three-dimensional flow features and heat transfer.

Kerrebrock et al. (2008) report on the blowdown testing of an aspirated counter-rotating fan, Figure 17.11. The basic idea is to control flow separation in axial compressors via the aspiration or suction of the viscous flows at diffusion-limited locations. To further increase stage work and pressure ratio while reducing weight and shortening, the compressor aspiration was

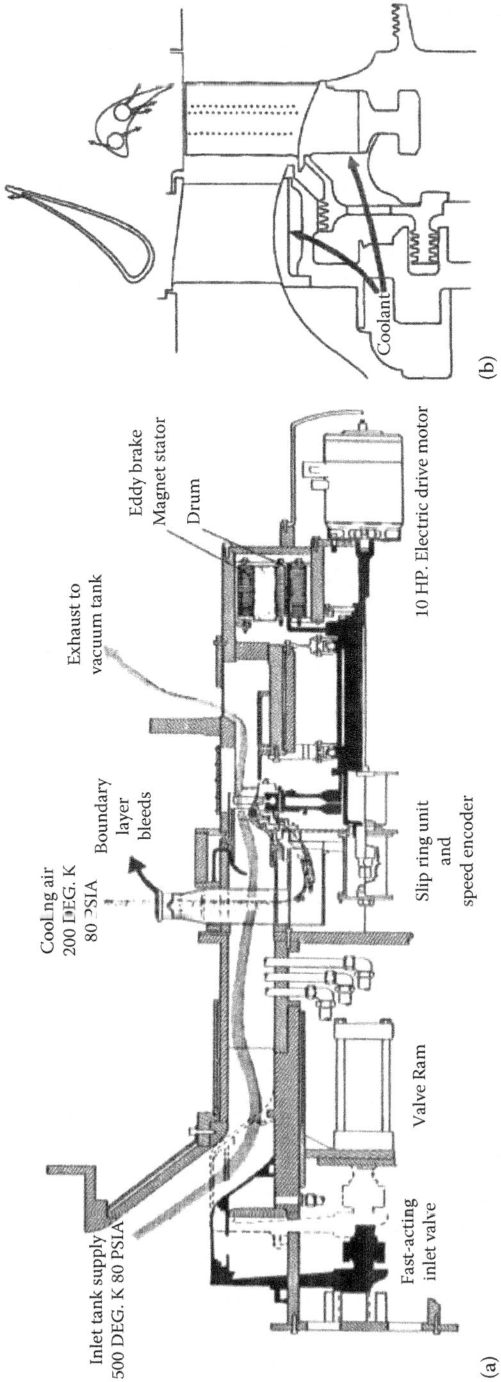

FIGURE 17.10
Turbine flow path and cooled turbine stage for a blowdown facility. (a) Schematic of moveable plenum wall and (b) detailed configuration.

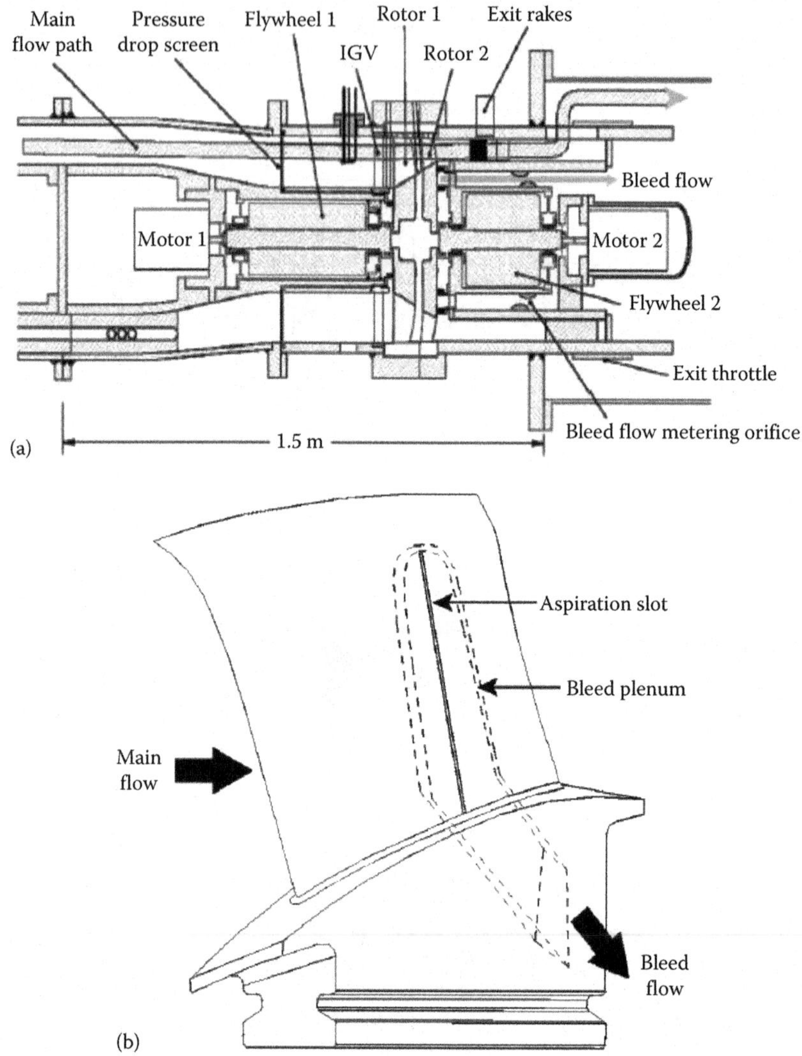

FIGURE 17.11
(a) Blowdown test rig with an aspirated counter-rotating fan stage and (b) aspirated rotor-2 detail.

combined with counter-rotation. In contrast to previous aspirated compressor work where shrouded rotors were used, the counter-rotating rotors were designed with conventional tip clearances. A stage pressure ratio of 3 was demonstrated at 87% isentropic efficiency.

Figure 17.12 shows a GTL test setup for shock tube testing of near-wall reaction effects on film-cooled heat transfer. Transient shock tube experiments were carried out to generate high-enthalpy flows (at 1000–2800 K

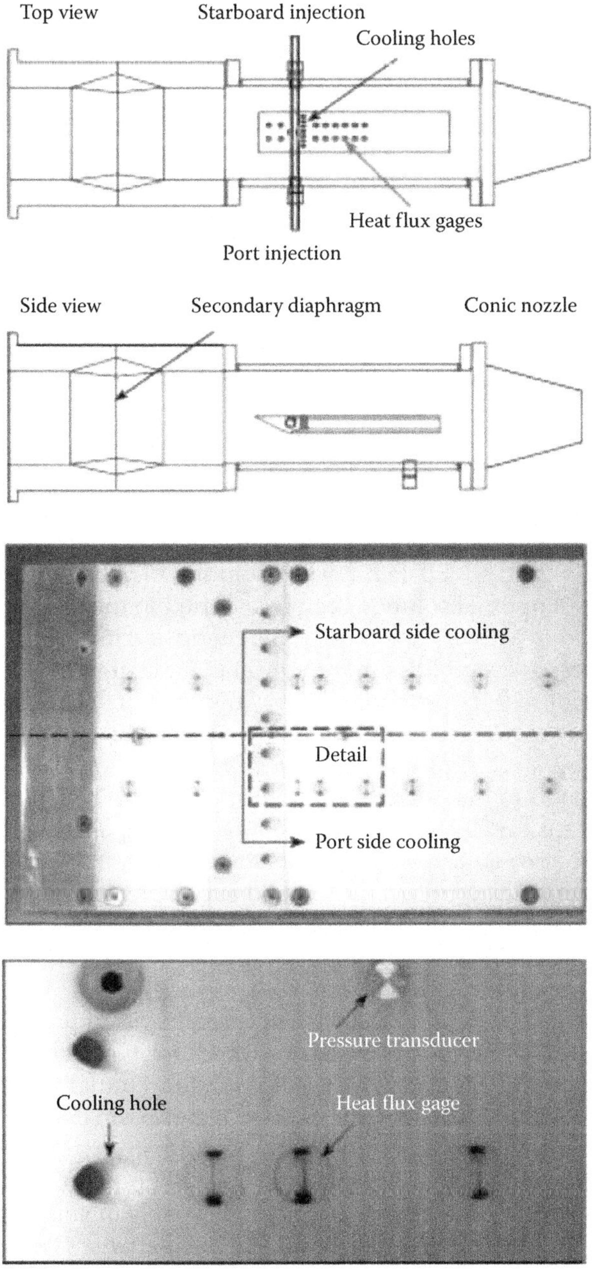

FIGURE 17.12
High-enthalpy transient shock tube testing of a film-cooled flat plate.

temperatures and at pressures of 6 atm) to assess secondary reactions near the surface of film-cooled flat plates. Ethylene–argon mixtures provided fuel-rich freestream flow, which reacted with the film-cooling air near the wall. Significant surface heat flux increases were measured, which can have implications on aero-performance and durability in commercial and military aircraft engines.

17.5 Smart Engines

Given advances in sensors, actuators, and microprocessors, concepts for *smart machinery/components* continue to proliferate, that is, *smart bearings, self-driving automobiles,* etc. In fact, the modern aircraft has been *pretty smart* for decades, in other words, it has an *automatic pilot* feature. Gas turbine jet engines are more or less open-loop devices, except that only the fuel control uses feedback. The idea behind *smart engines* is to change operation to closed loop control to improve performance and operability. An early concept for this was envisioned by Epstein (1986), pointing out that constraints can be relaxed and suboptimal solutions can be avoided in the design process. This idea is applied in active noise control, magnetic bearings, and active control of the compressor instabilities in rotating stall and surge.

Gysling et al. (1991) proposed a new method for centrifugal compressor surge line reduction by modifying the compression system dynamic behavior using structural feedback. One wall of a downstream volume is constructed so as to move in response to small perturbations in pressure to provide a means for absorbing the unsteady energy perturbations produced by the compressor so as to extend the stable operating range of the compression system. They utilized a lumped parameter analysis to define the coupled aerodynamic and structural system behavior and the potential for stabilization. Experiments were then conducted to examine the potential for stabilization concluded from the analyses.

The experiments validated the analysis predictions of roughly a 25% flow reduction of the surge line. Furthermore, because the tailored dynamics of the structure acted to suppress instabilities in their initial stages, the control was achieved with relatively little power being dissipated by the movable wall system and with no noticeable decrease in steady-state performance. Although the system was designed on the assumption of linear system dynamics, the structural control was shown capable of suppressing existing large-amplitude nonlinear limit-cycle surge oscillations. The test rig is illustrated in Figure 17.13.

Using a low-speed, single-stage, axial research compressor, Paduano et al. (1993) showed that the onset of rotating stall can be delayed in a low-speed, single-stage, axial research compressor using active feedback control.

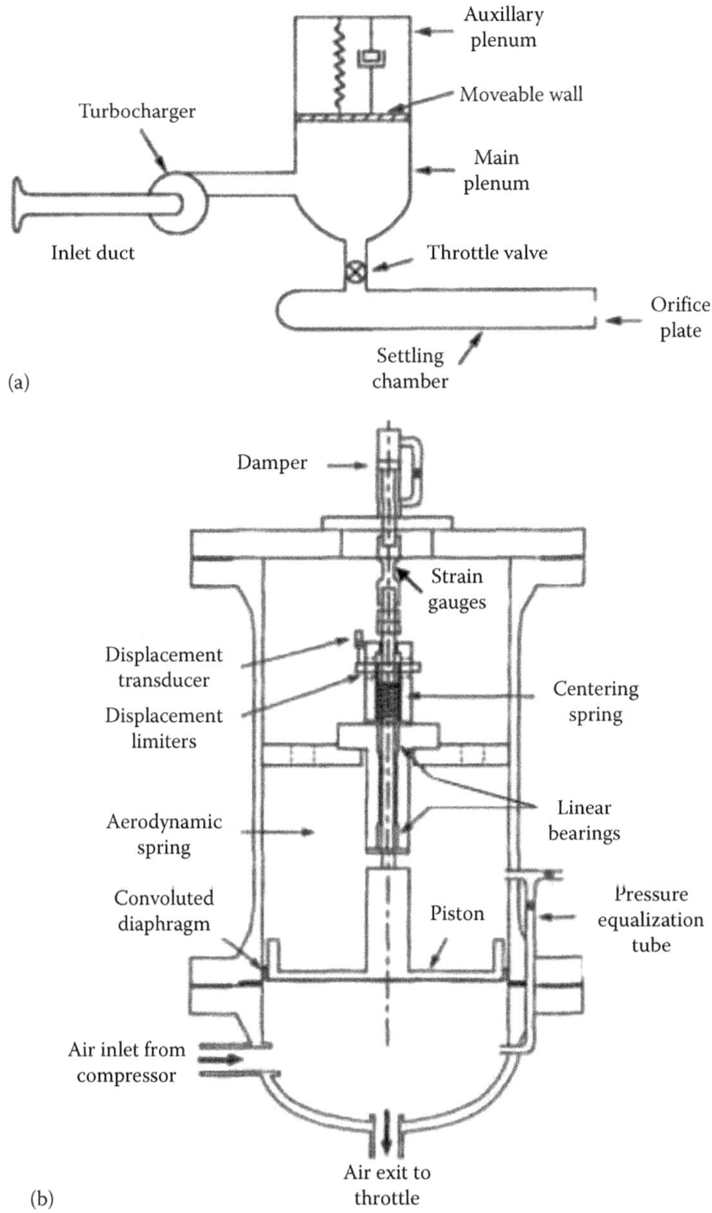

FIGURE 17.13
Movable plenum with mass-spring-damper system. (a) Schematic of moveable plenum wall and (b) detailed configuration.

Control was implemented using a circumferential array of hot wires to sense propagating waves of axial velocity upstream of the compressor. Using this information, additional circumferentially traveling waves were then generated with appropriate phase and amplitude by "wiggling" inlet guide vanes driven by individual actuators. The control scheme treated the wave pattern in terms of the individual spatial Fourier components with a simple proportional control law implemented for each harmonic. Control of the first spatial harmonic yielded an 11% decrease in the stalling mass flow, while control of the first, second, and third harmonics together reduced the stalling mass flow by 23%. The control system was also used to measure the sine wave response of the compressor, which exhibited behavior similar to that of a second-order system, thus providing a dynamic system identification of the compressor. The test rig for this research is illustrated in Figure 17.14. The author saw this test rig firsthand several years ago while visiting MIT to give an invitational lecture on rotor vibration to the MIT Mechanical Engineering Department's weekly Friday Colloquium.

Active control of rotating stall in multistage machines was also demonstrated in a three-stage axial compressor using high-speed control vanes between the IGV and the first rotor blade rows (Haynes 1993). The actuation is depicted in Figure 17.15.

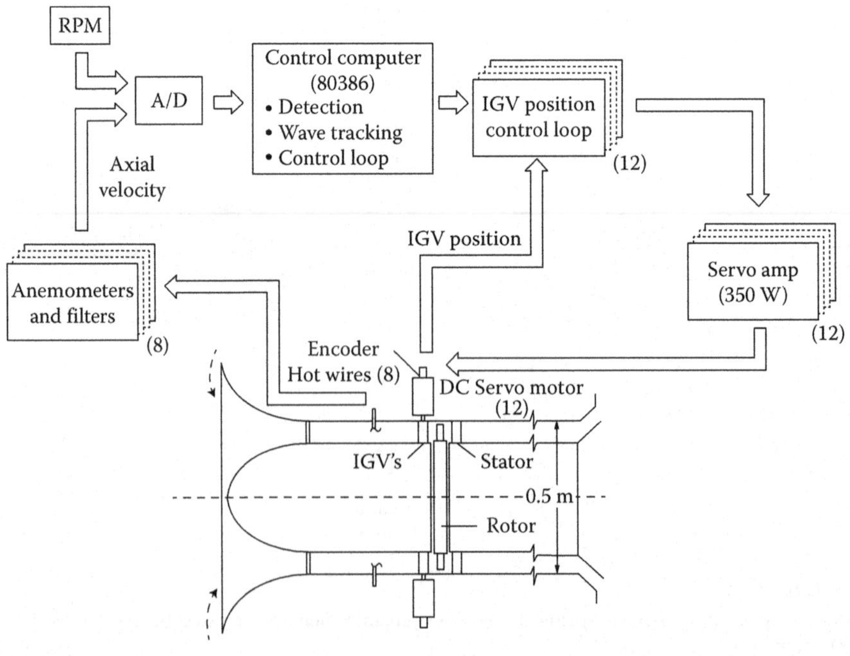

FIGURE 17.14
Active control of rotating stall in a single-stage compressor.

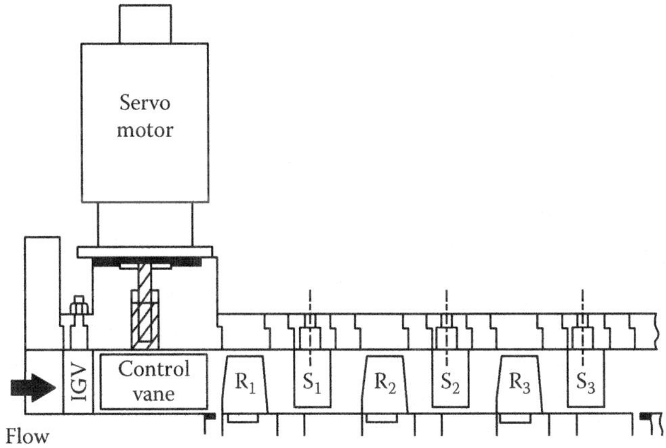

FIGURE 17.15
Three-stage compressor with individually activated high-speed control valves.

The structural control of rotating stall was demonstrated by Gysling and Greitzer (1995), who used an array of reed valves controlling the injection of high-pressure air in front of the compressor, Figure 17.16. The structural design chosen created the required phase relationships between the measured pressure and controlled flow rate.

Extensions to more complex situations with active control of rotating stall at the MIT GTL three-stage low-speed axial compressor include the following: (1) with inlet flow distortion (van Schalkwyk et al. 1998) and (2) with active control of transonic turbomachinery (Weigl et al. 1998). Active control

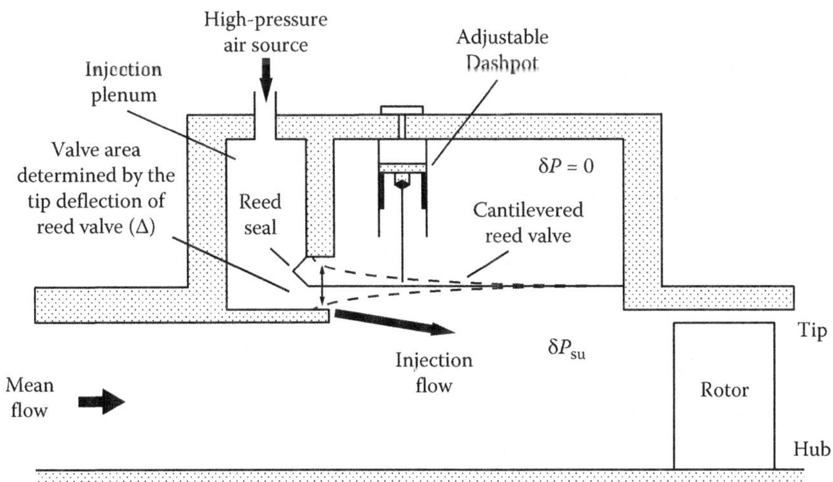

FIGURE 17.16
Schematic of dynamic mass/momentum injection.

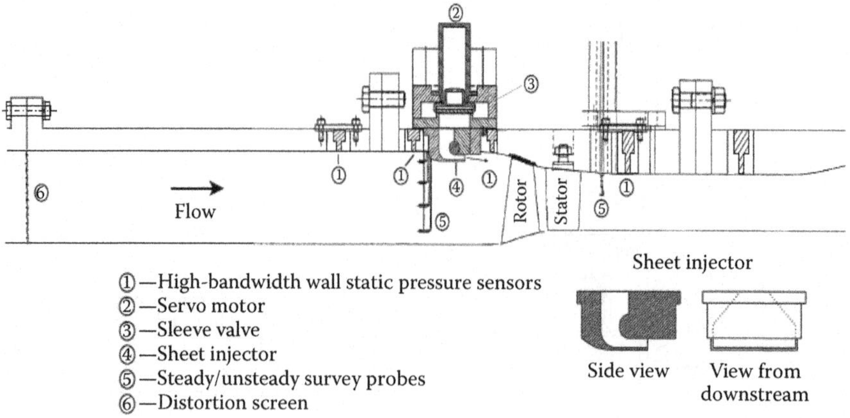

FIGURE 17.17
NASA Stage 35 actuation and instrumentation schematic with inlet distortion.

of rotating stall in NASA Stage 35 with radial and circumferential inlet distortion was demonstrated by Spakovszky et al. (1999), who determined that circumferential distortion leads to the coupling of modes between spatial harmonics. Figure 17.17 illustrates the NASA Stage 35 transonic axial compressor rig upstream with a distortion screen and circumferential array of sheet injectors.

Magnetic bearings have been explored for number-specialized rotor-bearing systems. They were even briefly considered by David Hibner of Pratt & Whitney (circa 1990s) for aircraft gas turbine double-spool shaft jet engine rotors. Although they offer some quite unique characteristics in rotor vibration control, their complexity, high cost, limited overload capacity, and the need for backup *drop down* bearings to prevent machine failure in case of an interruption of electric power to the bearings have all contributed to magnetic rotor bearings being relegated to the category of *novelty* (Adams 2010). Their use in one-of-a-kind research test rigs is not so inhibited.

However, work undertaken at the MIT GTL showed the potential application of magnetic bearings for high-speed compressor stall control. This non-rotor-support application is for active compressor blade-tip clearance control using magnetic bearings as servo-actuators to stabilize rotating stall. A conceptual design of an active stall control experiment with a magnetic bearing servo-actuator for the NASA-Glenn high-speed single-stage compressor test facility is reported by Spakovszky et al. (2001), Figure 17.18. Control laws developed to stabilize the compressor shaft, in a second control loop, employ a constant-gain controller to stabilize rotating stall. A detailed closed-loop simulation resulted in a reduction of stalling mass flow comparable to that obtained experimentally (Weigl et al. 1998), in the same compressor with air injection.

Active stabilization of surge was implemented on an Allied Signal LTS-101 axicentrifugal gas producer, reducing the surging mass flow by 1% for an operating range increase of 11% (Nelson et al. 2000). Control was achieved using high-response sensors in the inlet and diffuser throat, coupled to actuators that injected air near the diffuser throat, Figure 17.19. System identification and modeling revealed that a classical surge-type Eigen mode and an Eigen mode associated with engine duct acoustics dominated the engine's input–output interaction.

The surge mode's stability determines the open-loop surge mass flow. A robust linear controller with three inputs and one output stabilized this Eigen mode without destabilizing the acoustic mode. The controller facilitated a 1% reduction in surging mass flow at 95% N1 corrected. This increased the engine's choke-to-surge stable operating range by 11%. The measured unsteady presurge behavior of the engine and a systematic procedure for surge control law development are described by Nelson et al. (2000).

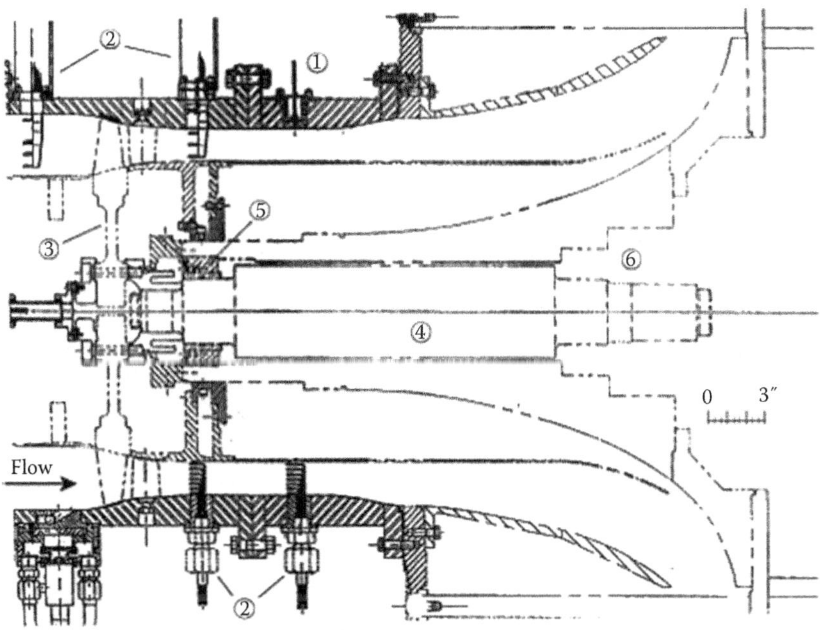

①—High-freq. static pressure sensors
②—Steady/unsteady survey probes
③—Compressor rotor disk
④—Compressor shaft
⑤—Front fluid journal bearing
⑥—Rear fluid journal and thrust bearing

(a)

FIGURE 17.18
NASA Glenn high-speed single-stage compressor facility and paper design of magnetic bearing servo-actuator for blade-tip clearance actuation: (a) sectional view of configuration. (*Continued*)

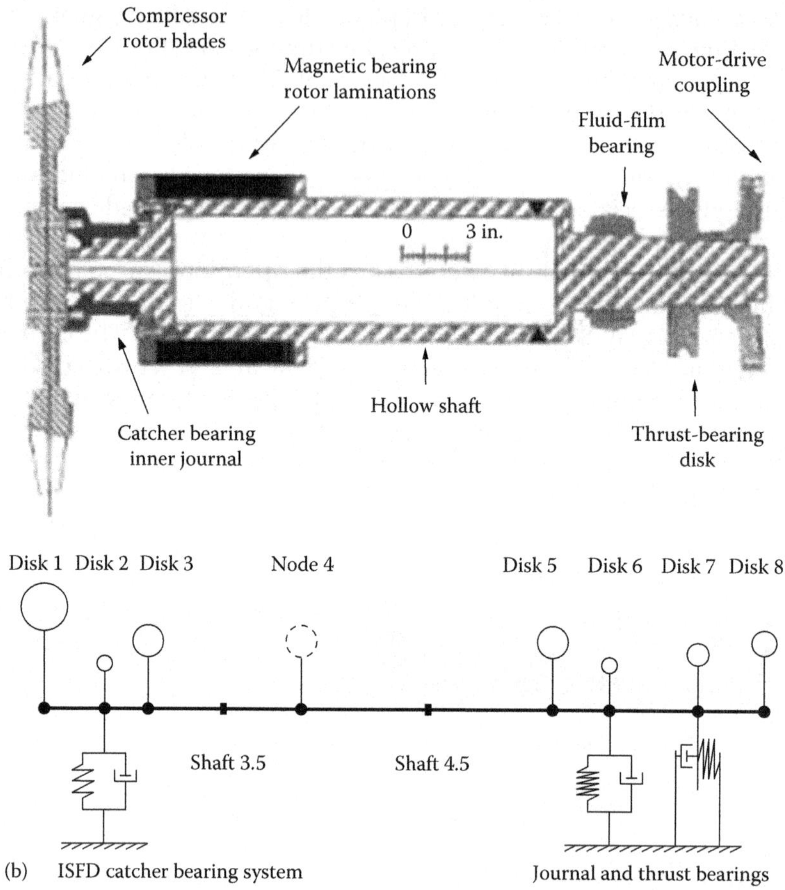

FIGURE 17.18 (*Continued*)
NASA Glenn high-speed single-stage compressor facility and paper design of magnetic bearing servo-actuator for blade-tip clearance actuation: (b) rotor vibration model.

17.6 Micro Engines

The basic objective of this project at GTL spanning more than 15 years was to shrink the gas turbine to millimeter scale to increase power density via the cube-square law by Epstein (2004), who provides a comprehensive list of references (74) and collaborators too numerous to mention here. Micro engines in power ratings from 10 to 100 W were made out of silicon wafers using computer chip manufacturing technology. These were MEMS-scale designs having centrifugal turbomachinery with pressure ratios in the range of 2:1–4:1 and turbine inlet temperatures of 1200–1600 K. Even with relatively low cycle efficiency, the increased power density suggests applications

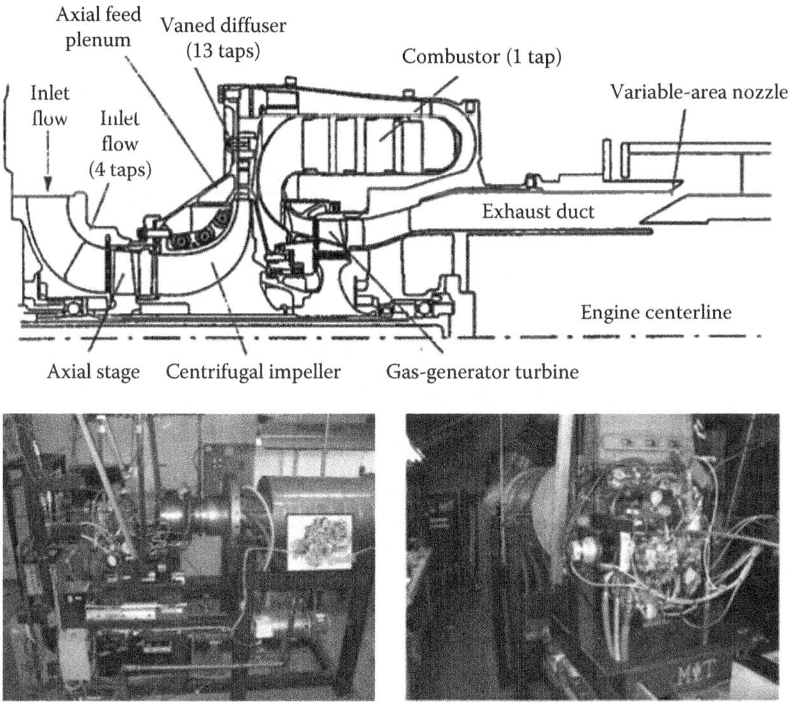

FIGURE 17.19
LTS-101 gas producer: note the plenum above the impeller shroud.

such as light-weight replacement of heavy batteries and propulsion systems for micro air vehicles. The key challenge is to achieve designs that meet thermodynamic and component functional requirements while economically produced with micromachining technology.

Millimeter-scale MEMS devices encompass micro gas turbine engines comprised of compressors, turbines, motors and generators, and micro rocket engines, including micro turbo-pumps and valves. Key developments of this work as reported by Lang (2009) include the demonstration of (1) stable combustion of gaseous and liquid fuels in microscale combustors, (2) low-Reynolds-number micro turbomachinery, (3) micro-size very-high-speed gas bearings, (4) micro-rocket-engine combustors, and (5) electrical and magnetic machines. Illustrations of micro-size turbomachine and component examples are shown in Figures 17.20 through 17.22.

More recently, Sato et al. (2011) reported on the design and testing of a palm-sized 50 W air-to-power (A2P) turbine generator powered by high-pressure air such as from a plant pneumatic system or portable bottle of pressurized air, 5–6 bars. It has an operating speed of 450,000 revolutions per minute (rpm) and a small blade span of 200 μm for optimum performance. Ceramic ball bearings were employed. The complete unit is a cylindrical device of

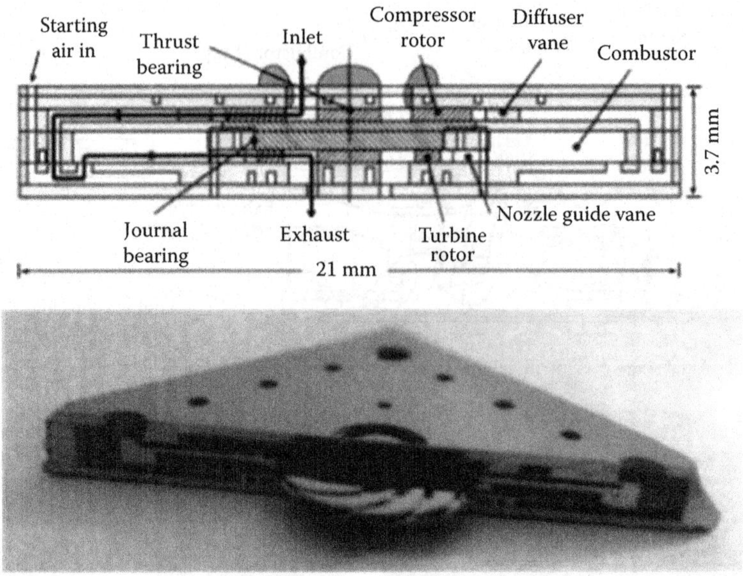

FIGURE 17.20
H_2 demo engine constructed from six silicon wafers (top), cutaway of H_2 demo engine (bottom).

FIGURE 17.21
4 mm diameter radial inflow turbine stage.

35 mm diameter resembling a tube fitting, Figure 17.23. Using load resistors, the proof-of-concept A2P device achieved 30 W of electrical power at 360,000 rpm and a turbine efficiency of 47%. The demonstrated performance was in good agreement with the computational fluid dynamics (CFD)-based predictions. Higher speeds under load could not be achieved due to thrust load limitations of the off-the-shelf ceramic ball bearings.

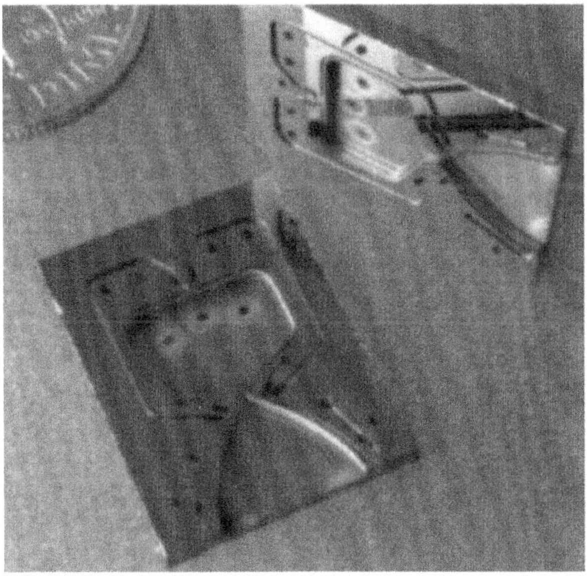

FIGURE 17.22
15 N thrust bipropellant liquid rocket engine.

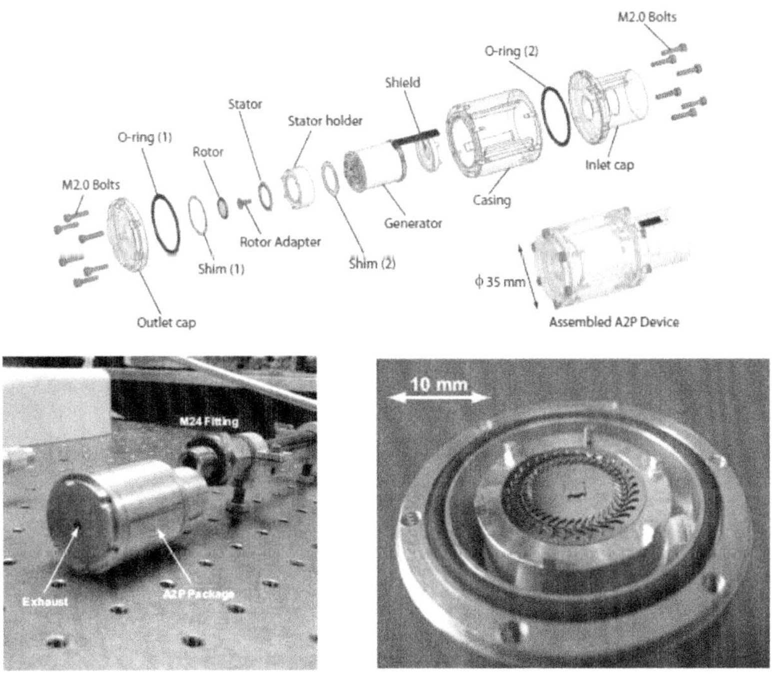

FIGURE 17.23
50 W air-to-power demonstrator.

17.7 Radial Turbomachinery Testing

The GTL transonic swirling-radial-flow generator test rig is illustrated in Figures 17.24 and 17.25. It was developed to study fluid dynamic phenomena in centrifugal compressor vaned and vaneless diffusers. Its unique feature is the capability of providing a wide range of diffuser inlet conditions, achieved via very-high-solidity rotating radial-outflow nozzle cascade combined with annular cross-flow injection suction slots upstream of the diffuser test section (Filipenco 1991). Mach numbers up to unity and flow angles of 63°–75° from the radial direction are achieved.

Filipenco et al. (2000) reported on a high-performance centrifugal compressor with radial discrete-passage diffusers. Two builds of discrete-passage diffuser were tested with 30 and 38 separate passages. Both the 30- and 38-passage diffusers exhibited a comparable range of installed operation and a similar level of overall diffuser pressure recovery. They concentrated on the influence of inlet flow conditions on the pressure recovery and operating range of radial diffusers for centrifugal compressor stages. The flow conditions examined include diffuser inlet Mach number, flow angle, blockage, and axial flow nonuniformity.

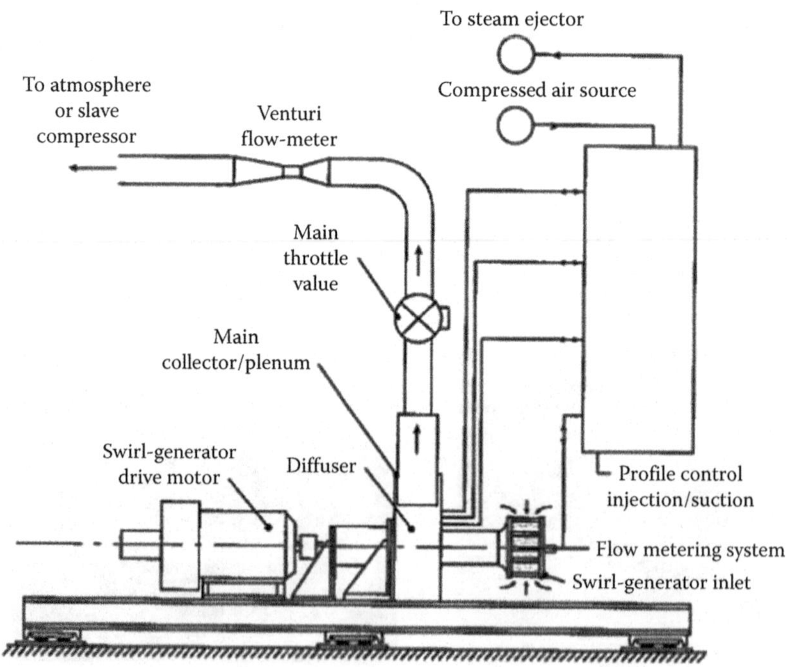

FIGURE 17.24
Overall test facility schematic.

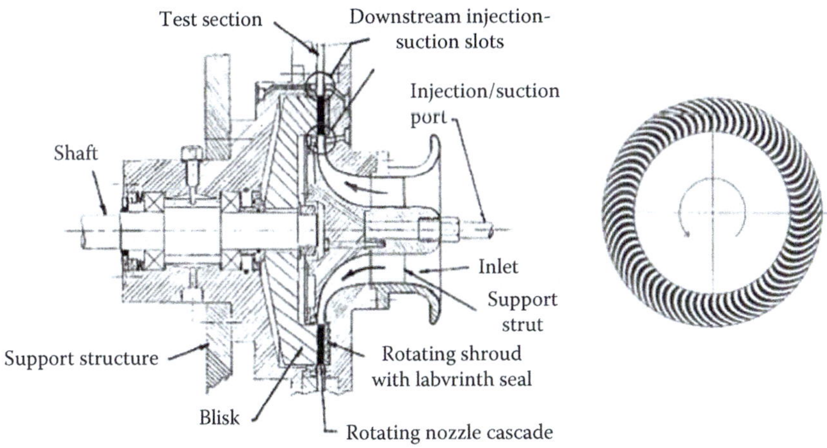

FIGURE 17.25
Swirling flow test rig detail with rotating-nozzle cascade.

A recent continuation of this work is reported by Everitt et al. (2016). The research focused on the effects of compressor impeller discharge conditions upon performance and stability with diffuser vanes. The modified swirl flow test rig is shown in Figure 17.26. A photo of the test rig is shown in Figure 17.27.

A major unknown is condensation effects in very-high-pressure ratio (~700) supercritical CO_2 compressors for Carbon Capture and Sequestration and Enhanced Oil Recovery. Phase change in turbomachinery has been studied extensively for steam turbines, where a significant amount of flow

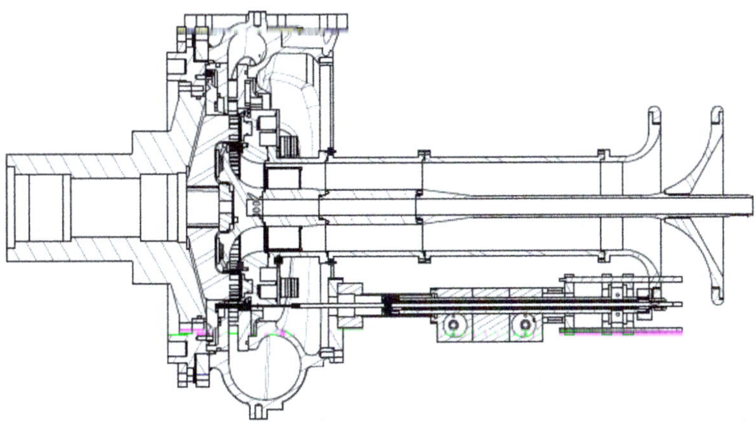

FIGURE 17.26
Cross section of modified swirl rig: increased span rotating-nozzle cascade, airfoil diffuser, and downstream volute and traverse system.

FIGURE 17.27
Test facility viewed from inlet end.

condenses as it expands through the last turbine stages (Gyarmathy [1962] 1964). Although this phenomenon is uncommon in compressors where the compression process occurs away from the two-phase region, there is a real concern and a debate in the CO_2 compressor community mostly because of the lack of data. Supercritical CO_2 real gas calculations have shown that local flow acceleration near the leading edge can lead to condensation pockets. Lettieri et al. (2015) established a new criterion for condensation onset in compressors. Computations and experiments demonstrate that condensation does not occur because the flow residence time typical in CO_2 compressors is much shorter than the nucleation time.

It is desirable to operate multistage, intercooled, super-high-pressure CO_2 compressors closer to the two-phase region to reduce stage inlet temperatures and therefore overall power requirement. In this, nonequilibrium phase change can occur, which poses major challenges in characterizing supercritical CO_2 compressor performance. While there is a large amount of literature focused on accurately characterizing the metastable state of metastable water vapor due to its prevalence as a working fluid in the power generation industry, there is limited information available for CO_2. Supercritical CO_2 blowdown experiments were conducted at GTL to measure metastable vapor properties at supercritical state using shearing interferometry. Experiments (Figure 17.28) verified metastable modeling assumptions and show good agreement with real gas computations (Paxon et al. 2016).

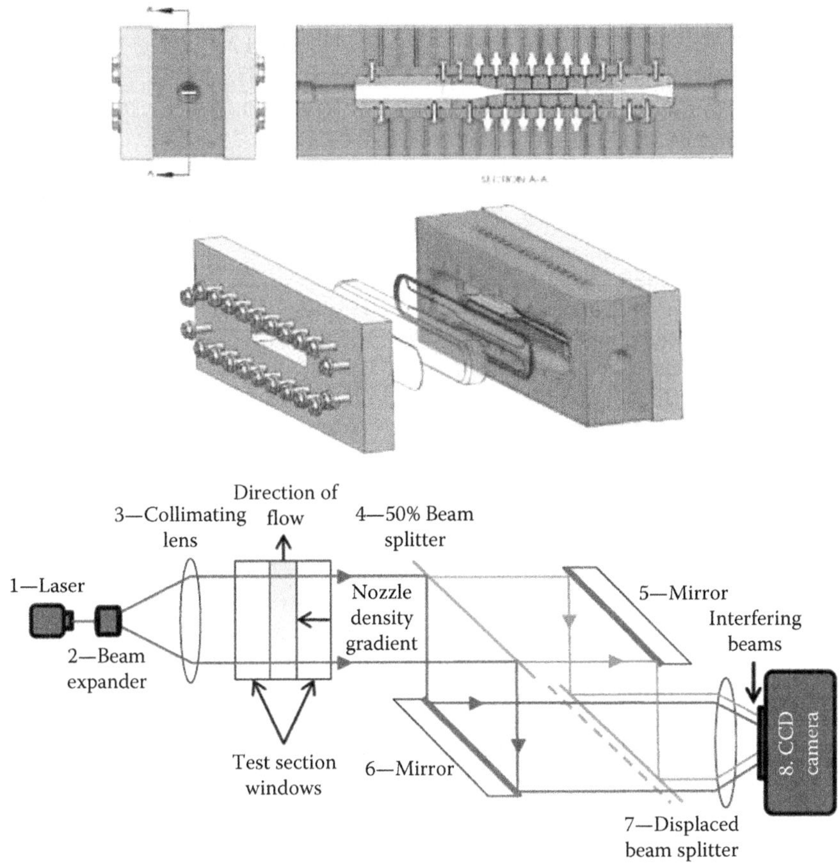

FIGURE 17.28
Supercritical CO_2 blowdown wind tunnel test section (top) and shearing interferometer (bottom) for testing at densities of up to 1000 kg/m^3.

Bibliography

Abhari, R. S. and Epstein, A. H., An experimental study of film cooling in a rotating transonic turbine, *ASME Journal of Turbomachinery*, 116, 63–70, 1994.

Adams, M. L., *Rotating Machinery Vibration: From Analysis to Troubleshooting*, 2nd edn., Taylor & Francis, CRC Press, Boca Raton, FL, 465pp., 2010.

Alford, J., Protecting turbomachinery from self-excited rotor whirl, *ASME Journal of Engineering for Power*, 87, 333–344, 1965.

Ehrich, F. F., Spakovszky, Z. S., Martinez-Sanchez, M., Song, S. J., Wisler, D. C., Storace, A. F., Shin, H.-W., and Beacher, B. F., Unsteady flow and whirl-inducing forces in axial-flow compressors: Part II—Analysis, *ASME Journal of Turbomachinery*, 123, 446–452, 2001.

Epstein, A. H., Guenette, G. R., and Norton, R. J. G., The design of the MIT blowdown turbine facility, GTL Report No. 183, Gas Turbine Laboratory, Massachusetts Institute of Technology, Cambridge, MA, 156pp., 1985.

Epstein, A. H., Smart engine components: A micro in every blade? *Aerospace America Magazine*, 24 (1), 60–64, Jan. 1986.

Epstein, A. H., Millimeter-scale, micro-electro mechanical systems gas turbine engines, *ASME Journal of Turbomachinery*, 126, 205–226, 2004.

Everitt, J., The role of impeller outflow conditions on the performance and stability of airfoil vaned radial diffusers, PhD thesis, MIT, Cambridge, MA, 285pp., 2014.

Everitt, J., Spakovszky, Z., Rusch, D., and Schiffmann, J., The role of impeller outflow conditions on the performance of vaned diffusers. *ASME Journal of Turbomachinery*, 2017, 139, 041004-1–041004-10.

Filipenco, V. G., Experimental investigation of flow distortion effects on the performance of radial discrete-passage diffusers, GTL Report No. 206, 99 373, Gas Turbine Laboratory, Cambridge, MA, 1991.

Filipenco, V. G., Deniz, S., Johnson, J. M., and Greitzer, E. M., Effects of inlet flow field conditions on the performance of centrifugal compressor diffusers: Part 1—Discrete passage diffuser, *ASME Journal of Turbomachinery*, 122, 1–10, 2000.

Gyarmathy, G., Grundlagen Eine Thsoeie Der Nassdampfturbine (Foundations of a theory of the wet-steam turbine), ETH, Zürich, Switzerland (English translation, 274pp., 1964, available for down load at: http://www.dtic.mil/get-tr-doc/pdf?Location=U2&doc=GetTRDoc.pdf&AD=AD0489324), 1962.

Gysling, D. L., Dugundji, J., Greitzer, E. M., and Eptein, A. H., Dynamic control of centrifugal compressor surge using tailored structures, *ASME Journal of Turbomachinery*, 113, 713–722, 1991.

Gysling, D. L. and Greitzer, E. M., Dynamic control of rotating stall in axial flow compressors using aeromechanical feedback, *ASME Journal of Turbomachinery*, 117, 307–319, 1995.

Haynes, J. M., Active control of rotating stall in a three-stage axial compressor, MS thesis, Mechanical Engineering, Massachusetts Institute of Technology, 217pp., 1993.

Hibner, D., Private communications with M. L. Adams, circa 1990s.

Kerrebrock, J. L., Epstein, A. H., Merchant, A., Guenette, G. R., Parker, D., Onnee, J.-F., Neumayer, F., Adamczyk, J. J., and Shabbir, A., Design and test of an aspirated counter-rotating fan, *ASME Journal of Turbomachinery*, 130, 1–8, 2008.

Kerrebrock, J. L., The M.I.T. blowdown compressor facility, GTL Report No. 108, 70pp., May 1972.

Lang, J. H. (Ed.), *Multi-Wafer Rotating MEMS Machines: Turbines, Generators, and Engines*, MEMS Reference Shelf, Springer Science & Business Media, New York, 453pp., 2009.

Lettieri, C., Yang, D., and Spakovszky, Z., An investigation of condensation effects in supercritical carbon dioxide compressors, *ASME Journal of Engineering for Gas Turbines and Power*, 137, 1–8, 2015.

Martinez-Sanchez, M., Jaroux, B., Song, S. J., and Yoo, S., Measurement of turbine blade-tip rotordynamic excitation forces, *ASME Journal of Turbomachinery*, 117, 385–392, 1995.

Millsaps, K. T. and Martinez-Sanchez, M., Dynamic forces from single gland labyrinth seals: Part II—Upstream coupling, *ASME Journal of Turbomachinery*, 116, 694–700, 1994.

Nelson, E. B., Paduano, J. D., and Epstein, A. H., Active stabilization of surge in an axicentrifugal turboshaft engine, *ASME Journal of Turbomachinery*, 122, 485–493, 2000.

Paduano, J. D., Epstein, A. H., Longley, J. P., Greitzer, E. M., and Guenette, G. R., Active control of rotating stall in a low-speed axial compressor, *ASME Journal of Turbomachinery*, 115, 48–56, 1993.

Paxon, D., Letteri, C., Spakovszky, Z., Bryanston-Cross, P., and Nakaniwa, A., Experimental assessment of thermodynamic properties for metastable CO_2, *Fifth International Symposium—Supercritical CO_2 Power Cycles*, San Antonio, TX, Mar. 28–31, 2016.

Sato, S., Jovanovic, S., Lang, J., and Spakovszky, Z., Demonstration of a palm-sized 30 W air-to-power turbine generator, *ASME Journal for Gas Turbines and Power*, 133, 1–10, 2011.

Shapiro, A. H., *Dynamics and Thermodynamics of Compressible Fluid Flow*, Reprint ed. with corrections, Krieger, Malabar, FL, 1983.

Spakovszky, Z., Analysis of aerodynamically induced whirling forces in axial flow compressors, *ASME Journal of Turbomachinery*, 122, 761–768, 2000.

Spakovszky, Z., Chapter 6: High-speed gas bearings for micro-turbomachinery, *Multi-Wafer Rotating MEMS Machines*, MEMS Reference Shelf, Springer Science & Business Media, New York, pp. 191–278, 2009.

Spakovszky, Z. S., Paduano, J. D., Larsonneur, R., Traxler, A., and Bright, M. M., Tip clearance actuation with magnetic bearings for high-speed compressor stall control, *ASME Journal of Turbomachinery*, 123, 464–472, 2001.

Spakovszky, Z. S., van Schalkwyk, M., Weigl, H. J., Paduano, H. D., Suder, K. L., and Bright, M. M., Rotating stall control in a high-speed stage with inlet distortion: Part II—Circumferential distortion, *ASME Journal of Turbomachinery*, 121, 517–524, 1999.

Storace, A. F., Wisler, D. C., Shin, H.-W., Beacher, B. F., Ehrich, F. F., Spakovszky, Z. S., Martinez-Sanchez, M., and Song, S. J., Unsteady flow and whirl-inducing forces in axial-flow compressors: Part I—Experiment, *ASME Journal of Turbomachinery*, 123, 433–445, 2001.

Thomas, H. J., Instabile Eigenschwingungen Turbinenläufern angefacht durch die Spaltströmungen Stopfbuschen und Beschauflungen (Unstable natural vibration of turbine rotors excited by the axial flow in stuffing boxes and blading), *Bull de L'AIM*, 71 (11/12), 1039–1063, 1958.

Thomas, H. J., Urlichs, K., and Wohlrab, R., Rotor instability in thermal turbomachines as a result of gap excitation, *VGB Kraftwerkstechnik*, 56 (6), 345–352, 1976 (in English).

Urlichs, K., Durch Spaltsrömungen Hervorgerufene Querkräfte an den Läufern Thermischer Turbomaschinen (Shearing forces caused by gap flow at the rotors of thermal turbomachines), Doctoral Dissertation, Technical University Munich, Munich, Germany, 1975.

Urlichs, K., Lekage flow in thermal turbo-machines as the origin of vibration-exciting lateral forces, NASA TT F-17409 (Die Spaltsrömungbei Thermichen Turbo-Maschinen als Ursache fürEnstehung Schwingungsanfachender Querkräfte), *Engr. Archiv*, 45 (3), 193–208, 1976.

van Schalkwyk, C. M., Paduano, J. D., Greitzer, E. M., and Epstein, A. H., Active stabilization of axial compressors with circumferential inlet distortion, *ASME Journal of Turbomachinery*, 120, 431–439, 1998.

Weigl, H., Paduano, J. F., Frechette, L. G., Epstein, A. H., and Greitzer, E. M., Active stabilization of rotating stall and surge in a transonic single stage axial compressor, *ASME Journal of Turbomachinery*, 120, 625–636, 1998.

Wohlrab, R., Experimentelle Ermittlung Spaltsströmungsbedingter Kräfte an Turbinenstufen und Deren Einfluss auf die Launfstabilitat Einfacher Rotoren (Experimental determination of forces conditioned by gap flow and their influences on running stability of simple rotors), Doctoral Dissertation, Technical University Munich, Munich, Germany, 1975.

Wright, D. V., Labyrinth seal forces on a whirling rotor, *Symposium on Rotor Dynamical Instability*, New York, *ASME Book*, AMD-Vol. 55, Adams, M. L., ed., ASME Applied Mechanics Division, University of Huston, TX, pp. 19–31, 1983.

18

TAMU Turbomachinery Laboratory

For more than the last 35 years, the Texas A & M University (TAMU) Turbomachinery Laboratory (TL) has conducted basic and applied research into important problems of reliability and performance of turbomachinery, which it succinctly defines as rotating machinery that extracts or adds energy to fluids. That's everything from classic Dutch windmills to the space shuttle's main engine turbopumps and industrial compressors that move natural gas through the distribution system. The TL's rotating machinery research prominence is widely recognized in the topics of (1) *fluid-film bearings*, (2) *seals*, and (3) *rotor dynamics*. This chapter covers TL research test rigs employed for these research topics. Publications of TL researchers utilized by the author in this chapter were recommended by the long time TL director, TAMU Professor Dara Childs.

18.1 Hybrid Hydrostatic–Hydrodynamic Journal Bearing

Kurtin et al. (1993) used a high-speed bearing test facility at the TL, Figure 18.1, to test a water-lubricated orifice-compensated 5-recess hybrid hydrostatic–hydrodynamic journal bearing under static load, for two radial clearance configurations. Measurements of journal-to-bearing eccentricity, torque, recess pressure, flow rate, and temperature were made at speeds from 10,000 to 25,000 rpm and supply pressures of 6.89 MPa (1000 psi), 5.52 MPa (800 psi), and 4.14 MPa (600 psi). For speeds of 10,000 and 17,500 rpm, the bearing load capacity was also "investigated." A pitching motion instability of the test bearing limited the number of test cases. Theoretical performance predictions were computed employing a 2D bulk flow model. Orifice discharge coefficients used in the computations were calculated from measured flow and pressure data. Reynolds number values for flow within the bearing lands due to shaft rotation and recess pressurization ranged from 6,700 to 16,500. The results of measurements and theoretical performance predictions were in reasonably good agreement.

The test rig illustrated in Figure 18.1 was also used by Childs and Hale (1994) to experimentally determine the stiffness, damping, and virtual-mass rotor dynamic coefficients plus the steady-state operating characteristics of the same high-speed hybrid hydrostatic–hydrodynamic journal bearing of

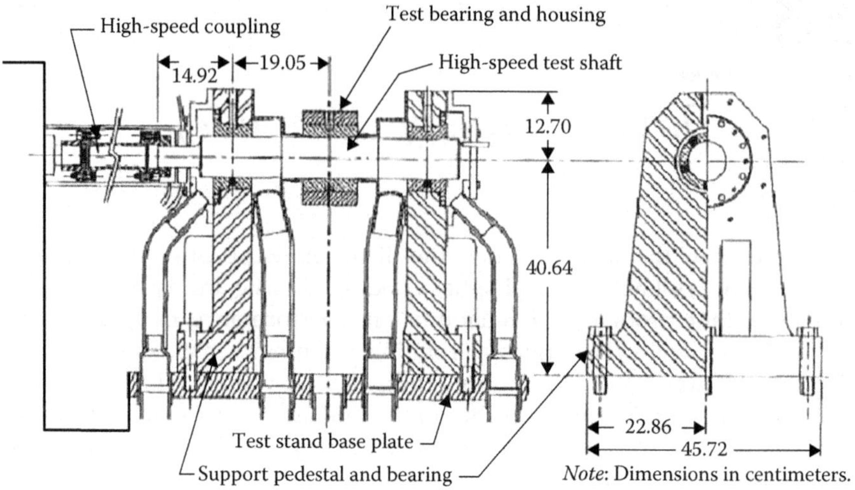

FIGURE 18.1
TL high-speed journal bearing test rig.

bore diameter 7.62 cm (3 in.). Speeds of up to the rig's maximum speed of 29,800 rpm were run with water as the lubricant in this series of tests, thus also achieving elevated Reynolds number test conditions. Static load on the bearing was independently controlled and measured.

18.2 Rotor Dynamic Coefficients of Plain Annular Seals

Marquette et al. (1997) developed reliable high-speed data for leakage and rotor dynamic coefficients of a plain annular seal at centered and radially eccentric positions using the TL high-speed journal seal test rig, Figure 18.2. A seal length/diameter ratio of 0.45 was tested, with the measured results having good signal-to-noise ratios. The influences on rotor dynamic coefficients of pressure drop, rotor speed, and static eccentricity were tested, with excellent agreement between experimental and theoretical results in the centered concentric position, in contrast to previous studies for the direct virtual-mass (inertia) coefficients. However, the agreement between experimental and theoretical rotor dynamic coefficients' results was not as good for radially eccentric runner positions. The authors therefore suggested that annular seals need to be modeled similar to journal bearings with rotor dynamic coefficients postulated for small motions about the static-equilibrium-position eccentricity, instead of postulating eccentricity-independent rotor dynamic coefficients. By then, the author had already, for quite some time, been occasionally suggesting that thought to Professor Dara Childs and was thus

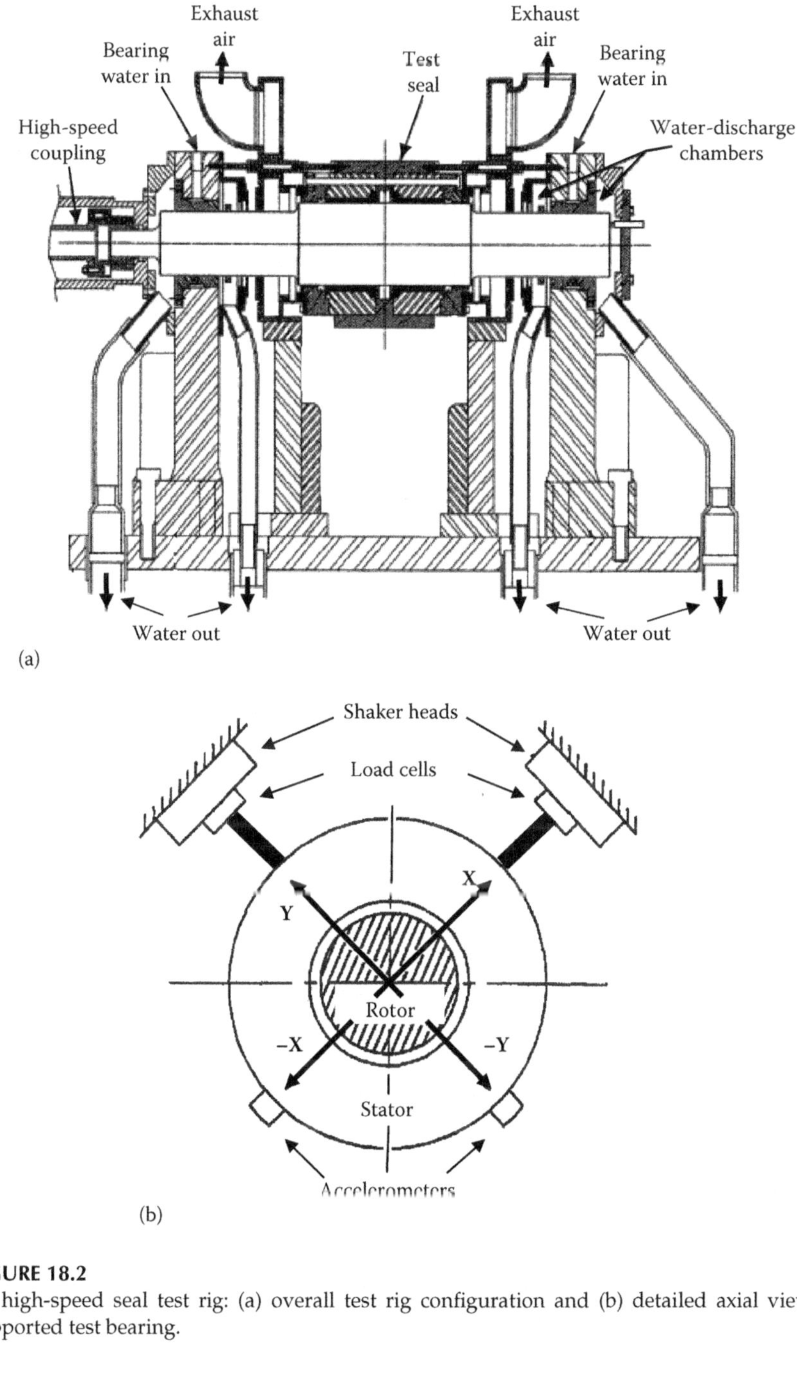

FIGURE 18.2
TL high-speed seal test rig: (a) overall test rig configuration and (b) detailed axial view of supported test bearing.

reassured when Dara finally came to the same conclusion in their 1997 paper. It's tough to argue with one's own carefully run test results.

18.3 Rotor Dynamic Coefficients of Tilting-Pad Journal Bearings

Tilting-pad journal bearings, also called pivoted-pad journal bearings (PPJB), are now used in a multitude of rotating machinery types, for example, turbines, compressors, pumps, motors, and generators. When properly designed and appropriately employed, this style of bearing yields distinct advantages in rotor vibration control over the purely cylindrical journal bearing. However, if improperly designed or misapplied, the PPJB can lead to undesirable results. A fundamental understanding of how a PPJB works can significantly increase one's potential in successfully utilizing it to its full advantages. To that end, Figure 18.3 illustrates fundamental differences between a PPJB and the purely cylindrical journal bearing (Adams 2010).

Rodriguez and Childs (2006) experimentally determined dynamic stiffness, damping, and inertia coefficients for a tilting-pad journal bearing. Their experimental results are compared to numerical predictions from models based on (1) the Reynolds lubrication equation for purely laminar viscous flow and (2) Navier-Stokes (NS) equations' bulk flow model that retains the temporal and convective fluid inertia terms. Their NS bulk flow model results correlate better with experimental results. The test rig used for this research is illustrated in Figure 18.4.

The bearing they tested was a 5-pad tilting-pad journal bearing, illustrated in Figure 18.5. Tilting-pad journal bearings typically have either a *cylindrical* or *spherical* pad–pivot contact element between each pad and the bearing surrounding pad-support ring, that is, either a *line contact* or a *point contact*. The additional advantage of the spherical point contact over the line contact is that it also makes the bearing pads axially *self-aligning*, for example, Figure 16.4(b). The bearing tested by Rodriguez and Childs used a somewhat newer approach to provide the required pad pitching degree of freedom. That is, with thin elastic ribs integrally connecting each pad to the outer bearing pad-support ring, as illustrated in Figure 18.5. This configuration has the advantages of (1) simplicity and is also (2) easier to install. But it has some disadvantages: (1) bearing pads are not axially self-aligning; (2) the pivot ribs have the potential for material fatigue failure, since the pads are naturally always in cyclic pitching motion to follow the ever-present residual rotor vibration; and (3) bearing pads do not allow adjustable preloading (Figure 18.3). Concerning this last disadvantage, the preload or lack thereof has to be precisely manufactured into the bearing bore and thus can't be

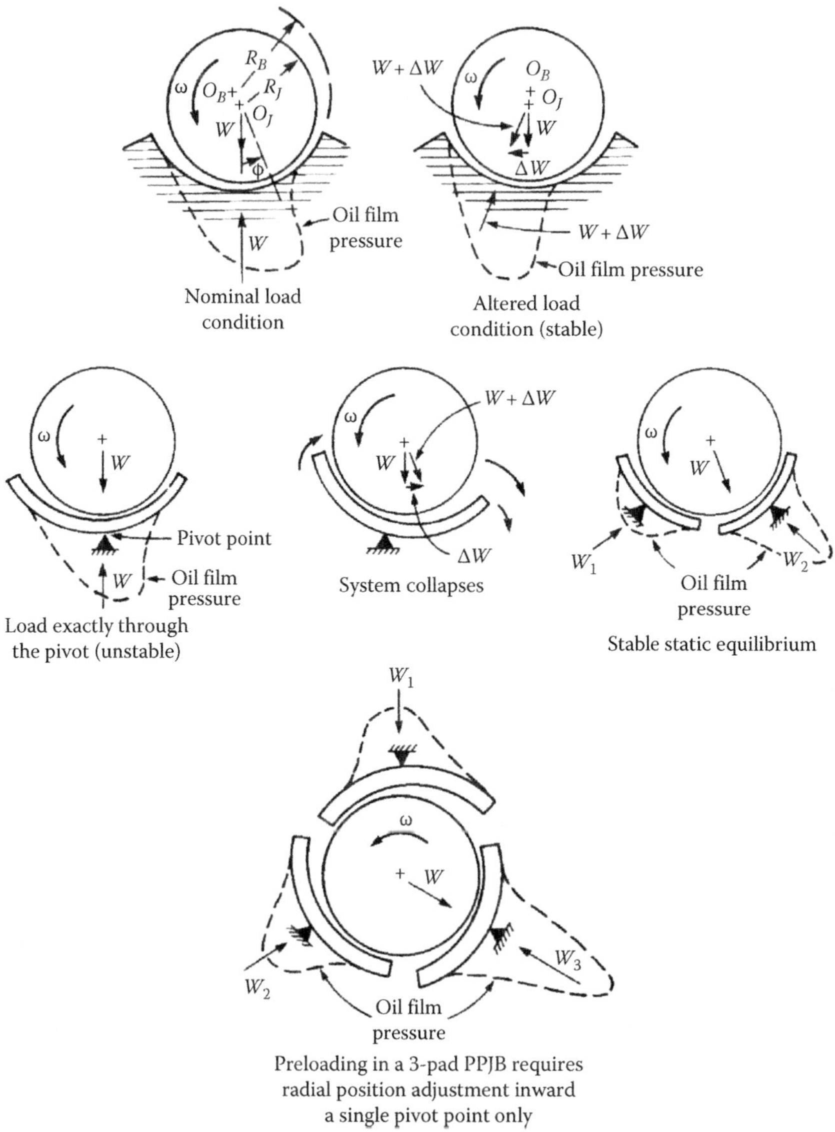

FIGURE 18.3
Comparison between cylindrical and pivoted-pad journal bearings.

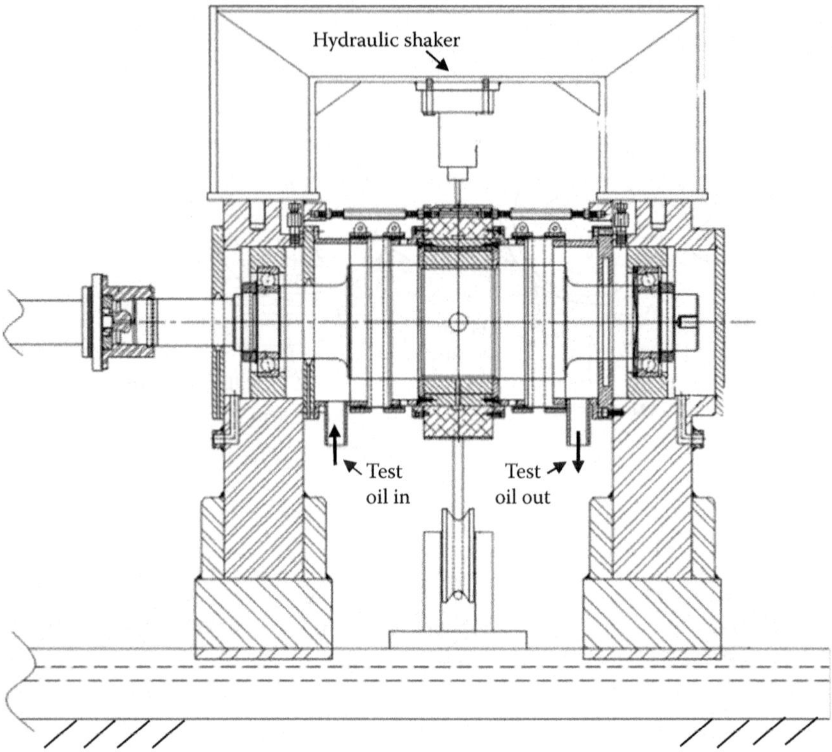

FIGURE 18.4
TL oil-lubricated test journal bearing rig.

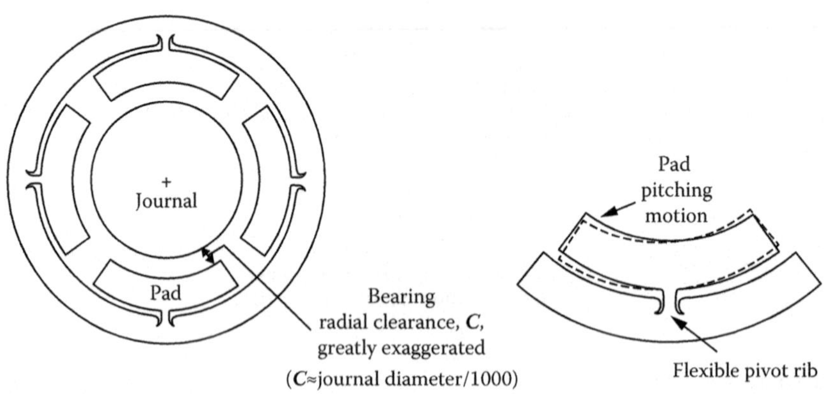

FIGURE 18.5
Tilting-pad journal bearing utilizing elastic flexible rib pivoting.

adjusted in the field, as can the more conventional *cylindrical* and *spherical* pad–pivot contact element configurations.

18.4 Honeycomb Gas Damper Seal

Honeycomb seals (Figure 18.6) are often considered as replacements for labyrinth seals in high-pressure centrifugal compressors to improve rotor dynamic stability. A practical concern naturally arose among centrifugal compressor designers that this enhanced stability characteristic of the honeycomb seal would deteriorate as the honeycomb cavities became clogged. Rodriguez and Childs (2006) conducted static and dynamic tests on a honeycomb seal and a smooth-bore seal, representing the honeycomb seal with completely clogged cells. Both were tested at the same constant clearances using air with a supply pressure of 70 bars. The test matrix included three speeds, three pressure ratios, and three inlet pre-swirl conditions. The test rig is shown in Figure 18.2. The results showed increased leakage, decreased synchronous stiffness, and decreased dynamic stability for the smooth seal with pre-swirled flow. The results strongly supported the use of swirl brake stationary vanes (Figure 17.4) at the entrance of a honeycomb seal if clogging is a concern. Comparisons between test results and predictions from a two-control-volume theory generally showed excellent agreement.

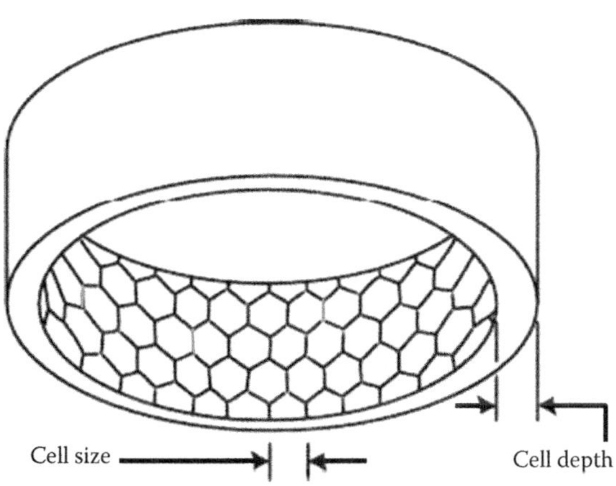

FIGURE 18.6
Honeycomb shaft seal for centrifugal compressors to increase rotor vibration damping capacity. (From Adams, M.L., *Rotating Machinery Vibration: From Analysis to Troubleshooting*, 2nd edn., Taylor & Francis, CRC Press, Boca Raton, FL, 2010, 465pp.)

Bibliography

Adams, M. L., *Rotating Machinery Vibration: From Analysis to Troubleshooting*, 2nd edn., Taylor & Francis, CRC Press, Boca Raton, FL, 465pp., 2010.

Childs, D. W. and Hale, K., A test apparatus and facility to identify rotordynamic coefficients of high-speed hydrostatic bearings, *ASME Journal of Tribology*, 116, 337–343, 1994.

Kurtin, K. A., Childs, D. W., San Andres, L., and Hale, K., Experimental versus theoretical characteristics of a high-speed hybrid (combination hydrostatic and hydrodynamic) bearing, *ASME Journal of Tribology*, 115, 160–168, 1993.

Marquette, O. R., Childs, D. W., and San Andres, L., Eccentricity effects on the rotordynamic coefficients of plain annular seals: Theory versus experiment, *ASME Journal of Tribology*, 119, 443–447, 1997.

Rodriguez, L. E. and Childs, D. W., Frequency dependency of measured and predicted rotordynamic coefficients for a load-on pad flexible-pivot tilting-pad bearing, *ASME Journal of Tribology*, 128, 388–395, 2006.

Sprowl, T. B. and Childs, D. W., A study of the effects of inlet preswirl on the dynamic coefficients of a straight-bore honeycomb gas damper seal, *ASME Journal for Gas Turbines and Power*, 129, 220–229, 2007.

19

University of Akron Bearing and Seal Lab

The University of Akron (UA) has had ongoing fluid-film bearing (FFB) research starting in 1977 with the author (1977–1982) and Professor Minel "Jack" Braun (1979–present). The author's main experimental FFB work while at UA is covered in Section 2.1, *squeeze-film dampers*. This chapter focuses on experimental research by UA Professor Braun, who recommended the publications on which this chapter is based.

19.1 Journal Bearing Oil-Film Rupture Visualization

Braun and Hendricks (1984) measured pressure and temperature in a Lucite bearing with static load imposed by fixing the journal-to-bearing eccentricity and tested at journal speeds of 2000–6000 rpm. Their results are presented in 3D plots and contour maps. Their flow visualization pictures are presented to show the downstream and upstream regions showing the region where the film ruptured evolving into the well-known finger striations, often referred to as *cavitation* although it is not the same phenomenon as *erosive cavitation* common in centrifugal pumps and hydro turbines. Pressure and temperature comparisons are made between tests with air-saturated oil and carbon-dioxide-saturated oil. Test results are compared with predictions using three theoretical models: (1) Swift-Stieber, (2) separation, and (3) Floberg. They assess each of these three models in light of their flow visualization test results. The test rig is illustrated in Figure 19.1.

Their test start-up from 0 to 2000 rpm was extremely fast, but the instruments and equipment used were intended to collect steady-state data rather than fast transients. However, a Fastax movie camera was used to record the incipient formation of the cavitation zone in both the upstream convergent recompression zone and downstream incipient nucleation zone portions on the journal. The first four consecutive frames at start-up and corresponding overlays are shown in Figure 19.2. They revealed the bubble in the process of developing from individual gas or vapor, which spreads in the oil from a cluster of bubbles and congregates to reach a critical radius and break the film homogeneity. As the four consecutive picture frames show, new small-bubble nuclei formed at the periphery of the main bubble and they are either engulfed in the larger gaseous structure or are carried downstream by the oil. This activity produced a rather unstable behavior of the cavity boundary.

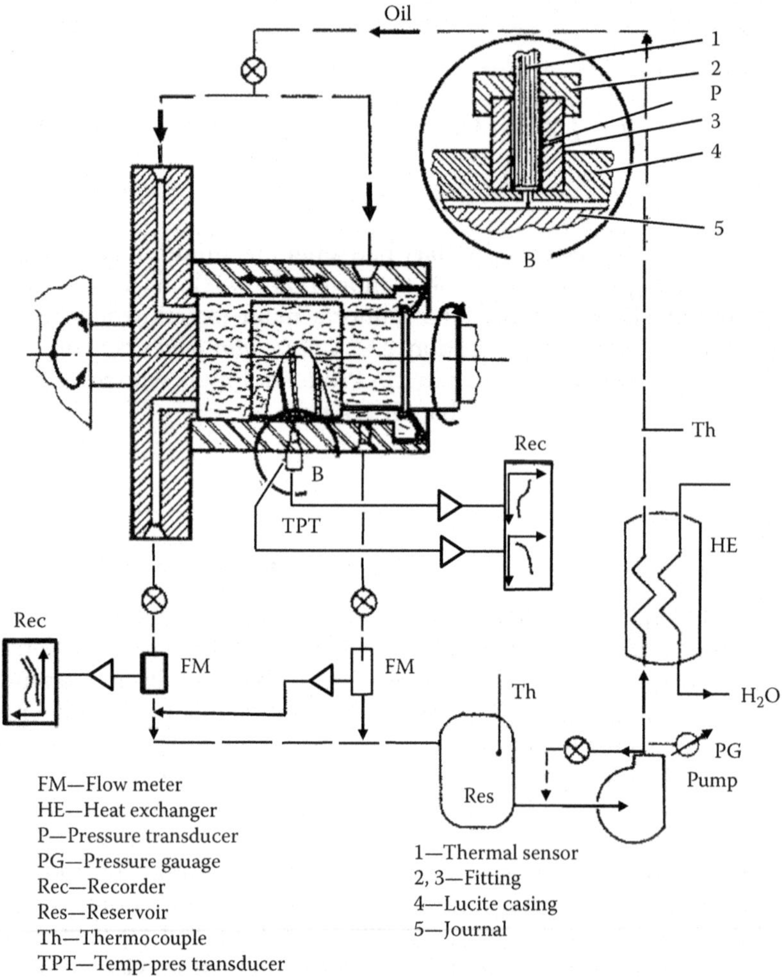

FIGURE 19.1
Journal bearing test rig with film flow visualization.

It became apparent that bubble formation was initiated at the point of lowest pressure around the minimum film thickness region when the journal reaches an angular speed at which the cavitation pressure is attained. Many of the bubble nuclei did not grow and were swept downstream. Braun and Hendricks characterized the development of the cavitation zone during this initial transient start-up stage as *pseudo-cavitation*.

As the journal velocity increased, the pressure dropped even further below the gas pressure in the vicinity of the minimum film thickness, while downstream regions drop to gas pressure. Thus, a sustained process of gas release occurred throughout the divergent section, enabling the bubbles to grow

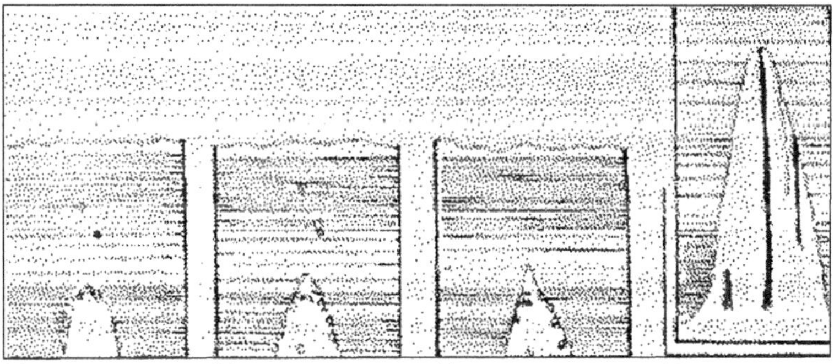

FIGURE 19.2
Cavitation inception and development during journal start-up: (a) beginning of speed up, (b) and (c) midrange speed up, and (d) at 2000 rpm.

through a process of mass addition and coalescence. The apex of the bubble extended into the film up to the point where the pressure in the oil film is greater than the gas pressure. Thus, the process of true gaseous cavitation replaced the pseudo-cavitation at the end of the speed transient to 2000 rpm.

19.2 Laser-Based Flow Measurements and Digital Image Processing

Braun and Hendricks (1991a,b) published a two-paper treatment (Parts 1 and 2) documenting this research. They used a technique based on computer-aided image processing with a full flow field tracking (FFFT) procedure to yield a nonintrusive method for the quantification of flow images. Their method yielded time-dependent trajectories, velocities, and accelerations throughout the entire flow field. In Part 1, they report on the application of the method applied to eccentric cylinders with the inner cylinder in rotation, that is, a journal bearing. The observed flow patterns were similar to the ones yielded by 3D computational fluid dynamics (CFD) simulations. The qualitative and quantitative parts of their study concentrated on the flow characteristics in and around the cylinders' point of closest approach, that is, minimum film thickness. The minimum film thickness was varied from 0.001 to 0.035 in. (0.0254 to 0.889 mm).

Their test rig and flow imaging setup are shown in Figure 19.3. They used a technique based on computer-aided image processing with an FFFT to yield a nonintrusive method for the quantification of flow images. Their method yielded time-set snapshots that revealed flow (Figure 19.4) as it enters at (a), emerges at (b), and out of the minimum film thickness zone at (c).

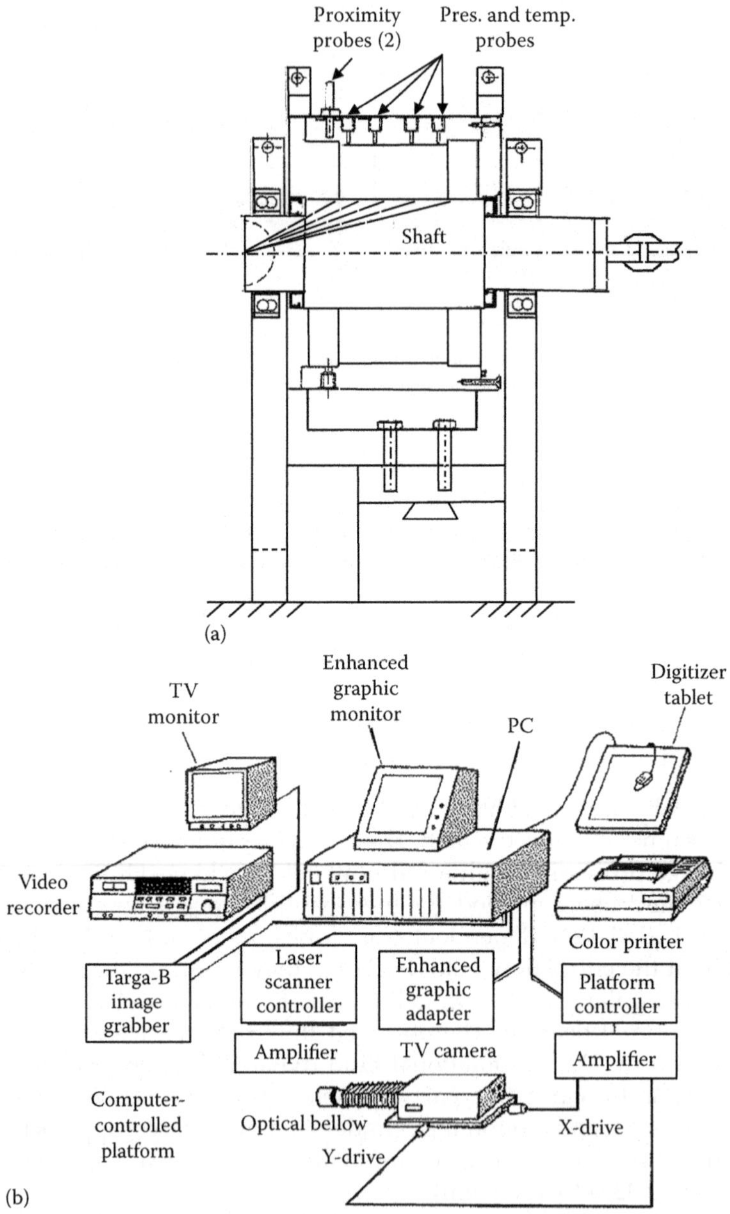

FIGURE 19.3
(a) Journal bearing rig and (b) flow imaging system.

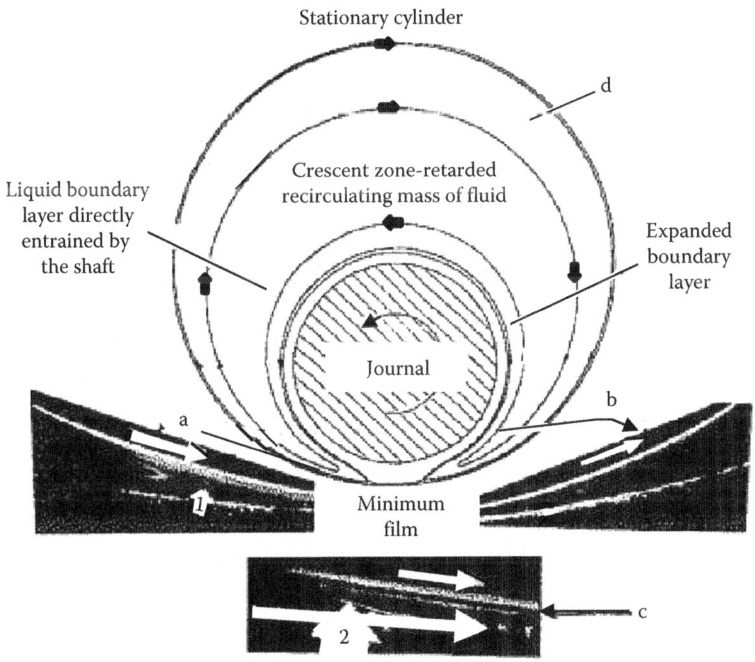

FIGURE 19.4
Sequence of photographs showing the flow pattern between the bearing and journal: (a) inlet convergent zone, (b) exit divergent zone, (c) minimum film, and (d) observed global flow pattern away from minimum film.

19.3 Hydrostatic Journal Bearing Flow Visualization

In their follow-on Part-2 paper, Braun and Hendricks (1991b) used the same facility shown in Figure 19.3 to study flow field visualization in a 6-pad hydrostatic journal bearing. Horvat and Braun (2011) compared experimental and CFD model predictions of flow inside the pockets of a 6-pad hydrostatic journal bearing. They thoroughly studied the flow patterns and pressure profiles inside both deep and shallow hydrostatic bearing pockets for ranges of pocket aspect ratio, restrictor resistance, and shaft surface velocity. For flow visualization, they used an FFFT method to track micro-sphere particles injected into the flow field to reconstruct flow patterns within the pockets. This is similar to the method used by Kadambi, as detailed in Chapter 6. Their test rig is shown in Figure 19.5 and flow visualization system in Figure 19.6. Moldovan et al. (2013) followed up with a 3D study with computational modeling and experimental flow visualization, also of a 6-pocket hydrostatic journal bearing.

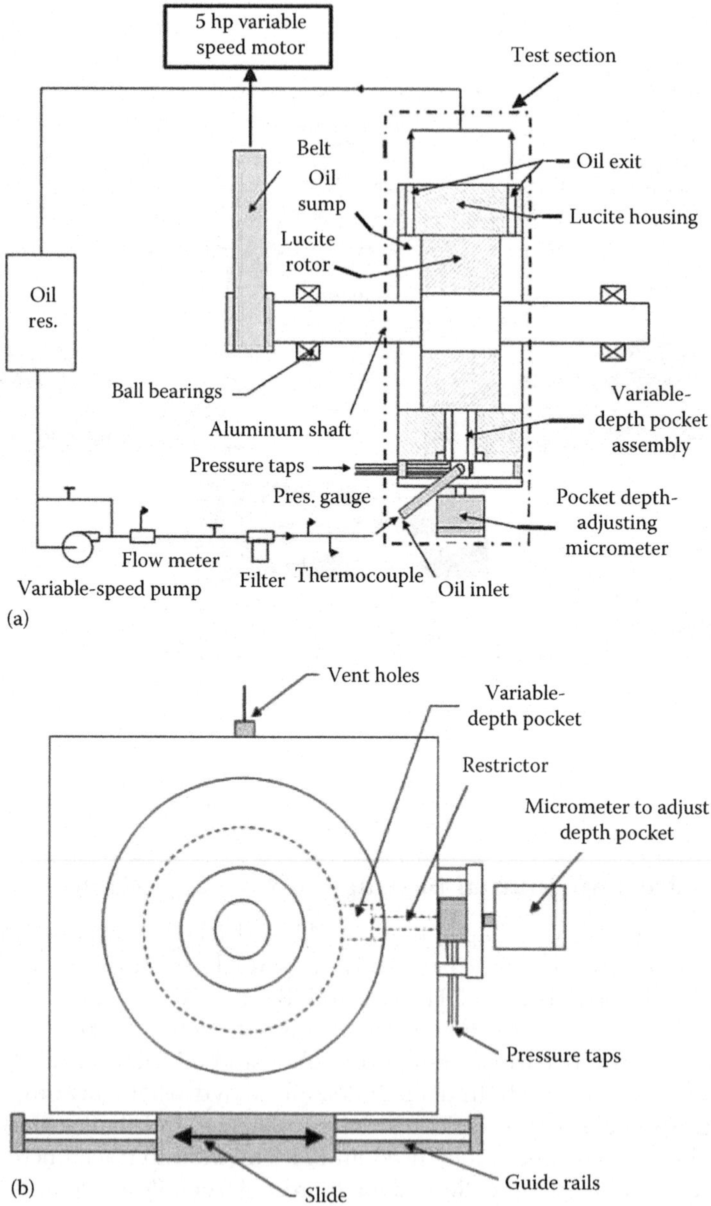

FIGURE 19.5
Schematic of test rig: (a) centerline view and (b) axial view.

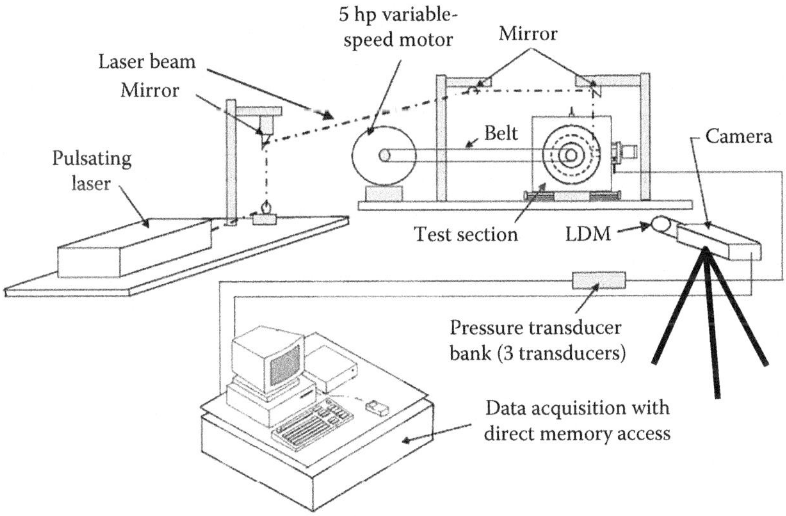

FIGURE 19.6
Sketch of vision system with long-distance microscope. *Note*: LDM, laser Doppler measurement.

19.4 Brush Seal Flow

Figure 19.7 illustrates a *brush seal*. In contrast to labyrinth, smooth-bore bushing, and Lomakin flat-tip toothed controlled-leakage radial seals, the rotating and nonrotating parts of a brush seal are in tight radially compressed contact, that is, with zero radial clearance. But that does not lead to seal seizure because the brush seal's flexible wire bristles are naturally compliant. The bristles, of course, also move radially compliant with the ever-present

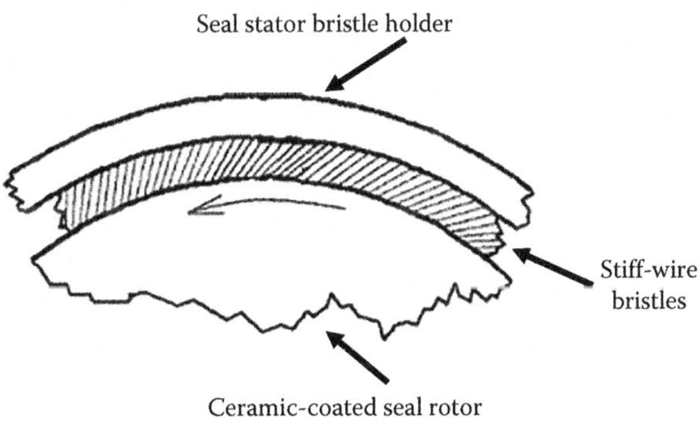

FIGURE 19.7
Brush seal.

residual rotor orbital vibration. With the significant radial compression of the bristles, the brush seal bristles can sustain a significant amount of wear and still remain radially compressed with zero radial clearance. As a consequence, brush seals are now used in high-tech applications involving high temperatures, high-pressure drops across the seals, and high rotor speeds, such as in modern gas turbines and aircraft jet engines. A NASA demonstration example 2-stage brush seal is pictured in Figure 19.8, alone and as installed in its demonstration tester.

Braun et al. (1990) designed a test tunnel to study fundamental flow patterns in a flow path with four in-series brush-seal-type barriers to flow.

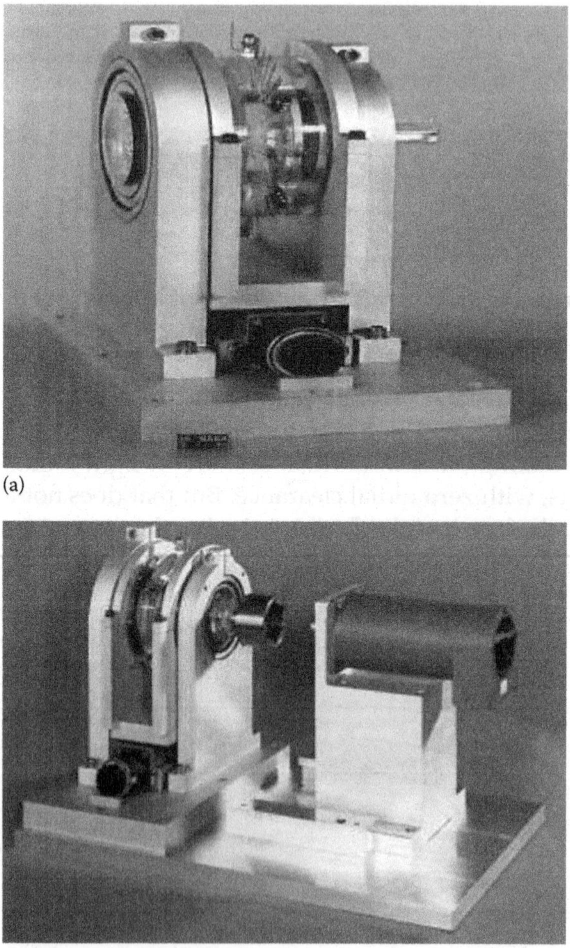

(a)

(b)

FIGURE 19.8
Tabletop demonstration brush seal (a) and as installed in a tester (b).

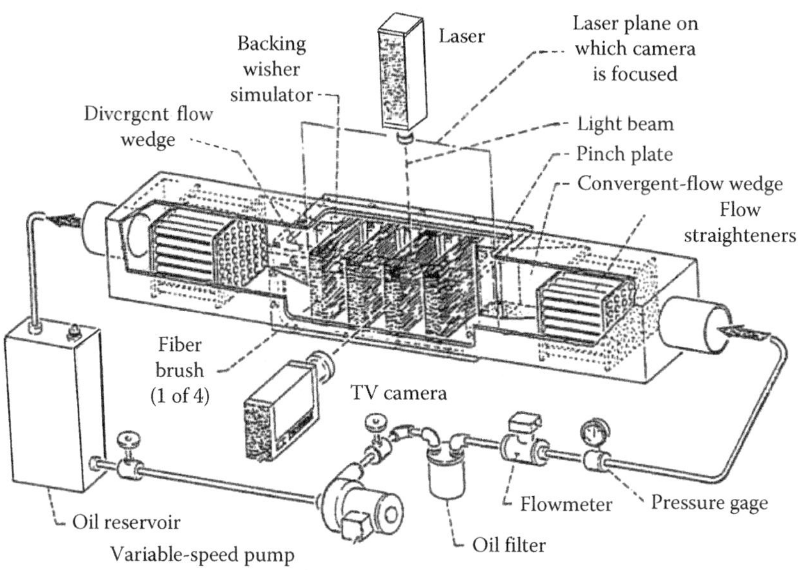

FIGURE 19.9
Test tunnel for investigating brush seal type flows.

This test apparatus is illustrated in Figure 19.9. Using FFFT, they nonintrusively produced graphically reconstructed flow visualizations. Their observed flow revealed combinations of river jetting and vortex patterns at locations upstream and downstream of the test section as well as in the zones between stages. Testing revealed flows that are highly sensitive to both spatial and temporal bristle voids, producing variation in seal leakage. The axial pressure variations across the four seal stages and seal leakage were measured as a function of seal pressure drop. Their further research using this facility is also reported by Braun and Hendricks (1991) and Braun et al. (1991).

Bibliography

Braun, M. J., Canocci, V. A., and Hendricks, R. C., Flow visualization and motion analysis for a series of four sequential brush seals, *Proceedings, AIAA/SAE/ASME/ASEE 26th Joint Propulsion Conference*, Orlando, FL, July 1990.

Braun, M. J. and Hendricks, R. C., An experimental investigation of the vaporous/gaseous cavity characteristics of an eccentric journal bearing, *ASLE Transactions*, 27 (1), 1–14, 1984.

Braun, M. J. and Hendricks, R. C., *Non-Intrusive Laser-Based Full-Field Quantitative Flow Measurements Aided by Digital Image Processing, Part-1: Eccentric Cylinders, Tribology International,* Butterworth Heinemann, Waltham, MA, Vol. 24, pp. 195–206, Aug. 1991a.

Braun, M. J. and Hendricks, R. C., *Non-Intrusive Laser-Based Full-Field Quantitative Flow Measurements Aided by Digital Image Processing, Part-2: The Hydrostatic Journal Bearing, Tribology International,* Butterworth Heinemann, Waltham, MA, Vol. 24, pp. 277–289, Oct. 1991b.

Braun, M. L., Canocci, V. A., and Hendricks, R. C., Flow visualization and quantitative velocity measurements in simulated single and double brush seals, *STLE Tribology Transactions,* 34 (1), 70–80, 1991a.

Braun, M. L., Hendricks, R. C., and Yang, Y., Effects of brush seal morphology on leakage and pressure drops, *Proceedings, AIAA/SAE/ASME/ASEE 27th Joint Propulsion Conference,* Sacramento, CA, June 1991b.

Horvat, F. E. and Braun, M. J., Comparative experimental and numerical analysis of flow and pressure fields inside deep and shallow pockets for a hydrostatic bearing, *STLE Tribology Transactions,* 54, 548–567, 2011.

Moldovan, S. I., Braun, M. J., and Balasoiu, A. M., A three-dimensional parametric study and numerical/experimental flow visualization of a six-pocket hydrostatic journal searing, *STLE Tribology Transactions,* 56 (1), 1–26, 2013.

Index

N

O

P

R

S